Biocultural Creatures

Biocultural Creatures
Toward a New Theory of the Human
Samantha Frost

Duke University Press   Durham and London   2016

© 2016 Duke University Press
All rights reserved
Printed in the United States of America on acid-free paper ∞
Typeset in Scala by Copperline

Library of Congress Cataloging-in-Publication Data
Names: Frost, Samantha, author.
Title: Biocultural creatures : toward a new theory of the human / Samantha Frost.
Description: Durham : Duke University Press, 2016. | Includes bibliographical references and index.
Identifiers: LCCN 2015044106|
ISBN 9780822361091 (hardcover : alk. paper) |
ISBN 9780822361282 (pbk. : alk. paper) |
ISBN 9780822374350 (e-book)
Subjects: LCSH: Human beings—Philosophy. | Human beings—Political aspects. | Philosophical anthropology.
Classification: LCC BD450.F767 2016 | DDC 128—dc23
LC record available at http://lccn.loc.gov/2015044106

Cover art: Marilene Oliver, *Split Petcetrix*, 2010.
Image courtesy of the artist.

For my mother, Dianne,
who taught me
how to be curious,
how to figure stuff out,
and how to carry on.

Contents

ix   Acknowledgments

1   **INTRODUCTION**

31   **CHAPTER 1** Carbon

53   **CHAPTER 2** Membranes

77   **CHAPTER 3** Proteins

101   **CHAPTER 4** Oxygen

119   **CHAPTER 5** Time

147   **CONCLUSION**

161   Notes
167   References
183   Index

Acknowledgments

First and foremost I want to thank the Andrew W. Mellon Foundation, whose provision of a New Directions Fellowship and a New Directions Post-Fellowship Award gave me the resources to train in a new discipline and to begin to come to terms with what I learned. This project simply would not have been possible without their generous support.

Thanks also to the University of Illinois Research Board for funds for research assistants and teaching releases, both of which made the research and writing processes a lot easier. I am also grateful to the Department of Gender and Women's Studies, the Department of Political Science, the Unit for Criticism and Interpretive Theory, and the College of Liberal Arts and Sciences for offering funds and awards that provided the conditions at critical moments for me to move forward with this project.

Gene Robinson graciously agreed to be my advisor as I began my course work. He has continued over the years to be a sounding board and a generous interlocutor as I tried to make theoretical sense of the world as seen on the figurative other side of campus. I am so very grateful for his gentle nudging and his careful, insightful questions.

Thank you—really, thank you—to the faculty and teaching assistants at the University of Illinois who opened their classrooms to me and taught me what they know: Tom Anastasio, Milan Bagchi, Bettina Francis, John Gerlt, Paul Gold, Jack Ikeda, Philip Janowicz, Donna Korol, Alejandro Lleras, Essie Meisami, Brad Merhtens, Jeff Moore, Mark Nelson, Kumar Prasanth, Michael Spies, Rebecca Stumpf, Fei Wang, and Maggie Wetzel. I am grateful to Darrell Kuykendall for reminding me how to solve "for" something in a formula without making me feel utterly foolish. Particular thanks go to Preston May, who met with me weekly in a nonvirtual face-to-face and who was patient, generous, and hopeful in coaching me through the trials and rigors of organic chemistry.

I am very grateful to my colleagues in Political Science, Gender and Women's Studies, and other places at UIUC. The support and friendship of Scott Althaus, Bill Bernhard, Jake Bowers, Jose Cheibub, Xinyuan Dai, Stephanie Foote, Pat Gill, Vicki Mahaffey, Colleen Murphy, Bob Pahre, Gisela Sin, Tracy Sulkin, Milan Svolik, Angeliki Tzanetou, Bonnie Weir, Matt Winters, and Cara Wong has been important throughout this project. Many thanks also to Michael Foellmer, Carol Hartman, Jacque Kahn, Brenda Stamm, and Virginia Swisher for their patience and skill in making the technical and bureaucratic dimensions of work proceed apace.

The inimitable Antoinette Burton pulled me aside after an early lecture for a provocative conversation, after which she became my dear reader—and eventually a dear friend. She read every chapter hot off the printer, and then the whole manuscript several times, always cheerful and enthusiastic, wryly pointing me in the right direction, poking at the knots, and taking pleasure in the unfolding of the project. Her generosity of spirit and acuity in engagement made it possible for me to keep going through to the end.

At various stages, my work on this project benefited from conversations with Nancy Abelmann, Jane Bennett, Jodi Byrd, J. B. Capino, William Connolly, Diana Coole, Charles Devellennes, Paul Duncum, Kennan Ferguson, Maria Gillombardo, Lauren Goodlad, Radhika Govindrajan, Susan Koshy, Craig Koslofsky, Sharon Krause, Trish Loughran, Justine Murison, David O'Brien, Bob Pahre, Dorothy Roberts, Michael Rothberg, Dede Ruggles, Roderick Wilson, and Linda Zerilli. I am also thankful for provocative discussions with members of the Biology of Social and Political Behavior group at the Institute for Genomic Biology, including Kate Clancy, Laura DeThorne, Ming Kuo, Tim Liao, Christy Lleras, Ruby Mendenhall, Vernita

Pearl Fort, Brent Roberts, and Gene Robinson. Hats off to Jesse Ribot, who read early versions of the first chapters and pushed me to think more politically about their import, and to Chris Larrison, who helped coordinate a seminar on the last chapter. I would also like to acknowledge the reviewers of my essay on biology and feminist theory, whose critical and probing questions helped me sharpen many of the arguments that wind through this book.

A huge thank you to my various research assistants who sought out sources, organized material, and conversed with me about my ideas: Jason Coronel, John Ostrowski, Karim Popatia, and particularly Mike Uhall, who worked through the final manuscript with me.

Thanks also to the graduate students whose stimulating conversation during seminars and other venues provided food for thought and an occasion for me to articulate what my argument would or should be: Claire Barber-Stetson, Noelle Belanger, Leslie Caughell, Patrick Fadely, Brandon Jones, Andrew Kaplan, Rebecah Jo Pulisfer, Wendy Truran, and Mike Uhall. A special note of recognition goes to Noelle for finding such an amazing cover image and for explaining it so beautifully to me. And of course, many thanks to Marilene Oliver for permission to use a reproduction of her artwork on the front of my book.

I had the opportunity to present the beginnings of my ideas to several audiences whose reception and response was helpful in pushing me to develop and articulate my ideas more clearly: The Matter Matters conference at University of Lund, the Matter, Life, Resistance conference at the University of Kent, the University of Virginia Political Theory Colloquium, the New Materialisms seminar at Rice University, and the Unit for Criticism and Interpretive Theory at the University of Illinois. The conversations with faculty and students at each of these venues were stimulating and clarifying.

Several people read the manuscript entire in an early form and offered me comments that were crucial to the final reshaping. Thanks to Stephanie Foote, Antoinette Burton, Andy Gaetke, Bruce Rosenstock, Chantal Nadeau, and Teena Gabrielson. The reviewers for Duke University Press provided exactly the kind of critical feedback that was needed to enable me to open up the text and make the book a better version of itself.

My editor Courtney Berger has been, as ever, a delight to work with. I am especially grateful for her confidence in the project and in my ability to bring it to fruition. Thanks also to Martha Ramsey for copyediting and to Sara Leone for shepherding the project through the production process.

Much appreciation goes to Jeff Melby, whose chiropractic wizardry throughout the past few years has quite literally kept me upright and forward-moving.

Early in the project, I was sustained by the friendship of Patricia Brady, Dan Kuchma, Annie McManus, Seth Mendelowitz, Steph Adams, and Brian Dill. As the project matured, Chantal Nadeau was a valuable interlocutor and a delightful companion. She inhabited the ideas with me, helped me wiggle the logic, and, perhaps more important, reminded me that sleeping, eating, and recreating are a necessity and not an indulgence. It is because of her insistence on the importance of play and pleasure that I am more vibrantly alive at the end of this project than I would have been otherwise.

I am lucky to have been able to lean on the encouragement and confidence of my mother, Dianne, and my sisters, Kim and Emma, as I worked. I am profoundly indebted to my children, Simon and Madeleine, for their support and understanding as I slogged my way through classes and writing. They offered tea, sacrificed bedtime reading, inquired after progress, and celebrated milestones as they came. I hope that I can be a better version of myself for them now that the project is done.

Introduction

Why a *New* Theory of the Human?

The aim of this book is to elaborate the basis for a new theory of the human. What provokes such a project is a concern that recent and withering critiques of the category of the human and of human exceptionalism have left us bereft of a politically useful category of the human subject that theorists can mobilize to address the political crises of the day. Under critical scrutiny, the notion of the human has come to be perceived as an index of a historically specific fantasy of mastery over the self, the earth, and all its many creatures. The characteristics, qualities, and capacities that heretofore have been taken to define and distinguish a human, humanity—*the* human—have been so profoundly discredited through historical, social, and scientific analysis that the notion itself seems to be bankrupt, with very little left to recommend it. In fact, given the weight of recent theoretical work challenging and discrediting not only the coherence of the category of the human but also the efficacy of human action, it may seem to be that to work to recuperate or reformulate the category is to kick against the pricks. Why bother?

At the same time that "the concept of the human has exploded," as

Rosi Braidotti puts it (2013: 1), humans are being called to task for their collective role in destroying the climate and the ecological resources of the planet and for their ineffectual social, political, and economic efforts to transform their modes of living in ways that are adequate to the gravity of the problem (Nixon 2011). As Steffen et al. (2011) explain, there is now almost incontrovertible evidence that "humankind, our own species, has become so large and active that it now rivals some of the great forces of nature in its impact on the functioning of the Earth system," so much so that "humankind has become a global geological force in its own right" (843). Indeed, surveying the "major and still growing impacts of human activities on earth and atmosphere, and at all, including global, scales," geologists along with climatologists and environmental scientists increasingly dub the era in which we live the Anthropocene (Crutzen and Stoermer 2000: 17). The likely disastrous climate and environmental transformations that are the mark of this age are a clarion call for humans to take concerted action to mitigate the effects of human exploitation of natural resources and the development of those resources into the means of our commodious existence. In other words, humans have to own up to and take collective responsibility for their role in precipitating crises of global proportion.

So, there is a bit of a theoretical problem in our understanding of what it is to be human. For as Dipesh Chakrabarty (2009) points out, even though the many critiques of the category of the human may incline us to deny that humans are anything as grand as "a geological agent . . . we appear to have become one at the level of species" (221). Moreover, in being called upon, urgently, to address the problems of climate change and environmental degradation, we are confronted with "a question of a human collectivity, an us" (222). If the convulsions and depradations characteristic of the Anthropocene demand that we think of humans as culpable and responsible for the current predicament, the inescapable question about the nature of the "we" implied in the question "what should or can we do" entails that we reconsider what it might mean to refer to, to invoke, or to try to mobilize a human subject.

The conviction that the idea of the human is not much more than a hollow fantasy coexists uneasily with the claim that humans as a species are a geological force. Of course, it might seem to be an abstruse theoretical conceit to advance the notion that the human is a fantasy at the same time that the very real effects of human industrial, agricultural, and consumer activity daily reshape the planet. But the strained antagonism between these two positions does not constitute an all-or-nothing impasse demanding that we throw up our theoretical hands in that ineffable gesture of dismissive exasperation. For a number of scholars have suggested that it is precisely because we have lived and labored under the fantasy of the human that we have wrought such terrible crises on the world (Bennett 2010; Latour 2004, 2013; Morton 2012). If this is the case, then a fundamental reconceptualization of what humans are, of what the human might be, could provide resources for cogent, creative, and robust engagement with the difficult question of how we should transform the ways we live.

As I shall show in the course of this introduction, many critiques of the idea of the human focus on the ways that the category is defined against a background of the natural and physical world, embodiment, animality, and organismic processes of living: these must be repudiated or denied value in order to construct the figure of an independent, self-sovereign, autonomous human agent. Implicit in the critiques that expose the fantasy as a grotesque and dangerous illusion are alternative figures for conceptualizing human creatures. These counter-concepts of the human are creatures who are embedded in various ecologies and networks of relations and who can integrate their acknowledgment of their embodiment, animality, physicality, dependence, and vulnerability into their self-conception and their orientation toward and modes of being in the world. This book is an effort to consolidate, animate, and give theoretical substance to these implicit counter-concepts. The hope is that if I can articulate and elaborate the theoretical shifts attendant to such an alternative, I might provide the basis for a conception of the human that can be theoretically serviceable and politically generative as we face the social and ecological devastation, the droughts, the famines, the population displacements, the economic and political upheavals, and the wars that apparently accompany the inexorable unfolding of the Anthropocene.

In transforming the chastening critiques of the human into an affirmative alternative, the book proposes that we conceive of humans as biocultural creatures. Exactly what that might mean will hopefully become clear in the pages and chapters that follow. But a brief orienting overview here of these key terms will perhaps set the stage for engagement.

So, I use the term "creatures." To describe humans as creatures here is to be held to account for human creatureliness, for the ways that humans, like all other creatures, are alive and are able to stay alive because they are embedded in and draw manifold forms of sustenance from a habitat of some kind. To insist on human creatureliness is not to deny humans' difference from other creatures—all creatures are different from one another, and there does not seem to be a particularly good reason to refuse to acknowledge the difference of human creatures from other creatures too. Rather, to insist on human creatureliness is to refuse the hubristic exception that would make humans a bizarre and almost unthinkable living phenomenon, abstracted from the habitats that are the condition of their being able to live.

I also use the term "biocultural." All creatures are biocultural in the sense that they develop, grow, persist, and die in an environment or habitat that is the condition for their development, growth, persistence, and death. I settle on the term "biocultural" as a descriptor for humans because I want to create a conceptual binding or a constraint such that we can no longer disavow what has been most vehemently disavowed—our biological, organismic, living animality. And rather than using a term like "human animal" to capture this, I think "biocultural" encapsulates the mutual constitution of body and environment, of biology and habitat that has been so central to the challenge to the category of the human.

The "-cultural" part of the term is used here with the sense not of a noun but of a verb, as in "to culture," to cultivate, to provide some kind of medium within which a thing or things can grow. I prefer to think of culture in terms of the verb because it nudges us to take into consideration not just dimensions of our living habitats that shape and give meaning to living bodies and deeply complex forms of social and political subjectivity but also those dimensions that materially compose living bodies. In other words, to think of culture in terms of cultivation enables us to incorporate in our thinking simultaneously and sometimes variously aesthetic, arboreal, commodified, composted, cruciferous, disruptive, economic, embodied, fermented, fractured, institutional, linguistic, metaphoric, normative, oceanic, organic, patterned, political, representational, subjective, trashed, treasured, violent . . . in short, the material, social, and symbolic worlds we inhabit. All these dimensions of the medium in which creatures are cultured are important to take into account.

Figuring out how to think about the "bio-" in "biocultural" is tricky. It is tricky and difficult, I think, not necessarily because of disciplinary and subdisciplinary overspecialization, as Goodman and Leatherman (1998) propose, but rather because our conceptual habits and our philosophical vocabulary and grammar inevitably draw into the ambit of our thinking some notion of a pure, unadulterated, uncultivated dimension of a living body. Later in the book, I talk about this imaginary sequestered, uncultivated piece or aspect of living bodies as a fictive "überbiological" matter—something that persists unaffected by and in spite of environmental insult. In appending "bio-" to "-cultural" to form a working concept of "biocultural," I push against this unwitting tendency to presume the "überbiological" and instead try to capture and elucidate how creatures become what they distinctively are through the habitats that culture them.

In the remainder of the introduction I will trace the contours of the debate that critically engage the idea of the human. In doing so, I will provide a theoretical context for the idea of the biocultural that is the focus of the chapters that follow.

The Critique of the Human

Critics of the idea and category of the human pursue three interrelated approaches, challenging the composition of the group that comprises human individuals, questioning the purview of rights that attach to the human as a moral and political category, and contesting what is distinctive or characteristic in the actions that human agents undertake. The object of their critique is the following set of ideas: that to be human is to be a member of a moral community composed of those who consider themselves as and treat one another as equals. That to be human is to possess all the rights and responsibilities implied by and presumed to follow from that equality. And that, accordingly, the concept of the human implies moral worth and carries ethical duties or obligations for those captured under its umbrella. In fact, Jürgen Habermas (2003) explicitly ties together the social, the moral, and the biological, arguing that the idea of equality and the moral and political community that such equality makes possible depend crucially on a presumption that there is an essence of human lodged in a necessarily inviolable biological strata. Defending the idea of the human, Habermas (2003) claims that it is important to preserve some notion of a distinctive human nature because the moral duties and obligations that attend the idea of that distinctiveness provide the model for all other forms of moral limits and duties. But as I will show, critics of the idea and category of the human are considerably less convinced that "the human" is a guarantor of a right and just collective life.

As an ideal, the category of the human has been construed as universal, which is to say that everyone supposedly is captured under its rubric. However, historically, those people considered to fall properly within the domain of the human have been constricted by religious, social, political, or cultural mores. As numerous scholars have shown, women, the laboring classes, the physically and mentally disabled, queers, and people of various cultural or national origins have been construed as dubiously human—as almost-but-not-quite human—because they have been perceived as lacking the clarity of mind and dispassionate distance in judgment afforded by the gift of reason (Baynton 2001; Bordo 1987; Chen 2012; Irigaray 1985; Lloyd 1984; Luciano and Chen 2015; Macpherson 1962; Mills 1997; Okin 1979; Pateman 1988; Schiebinger 1993). Distinctively trapped in and by the body, these "others" of the fully human were construed both as subject to the determinations of the biological or animal functions of the body and as vulnerable to a kind of behavioral determinism deriving from the failure of their "weak" intellects to isolate them from the solicitations, seductions, and predations of the social and cultural milieux. This putative intellectual and moral weakness meant that they could not be received as the equals of those who counted themselves rational. Lacking the capacity for self-mastery, they had to be managed and spoken for or represented by their supposed betters. As a consequence of being inferior versions of the human—indeed, if they were presumed to be human at all—they were not entitled to claim the rights or exercise the responsibilities that accrued to those who were fully human. Indeed, the subordination that seemed to arise from their very nature was often seen as both an explanation and a justification for their being subject to the control, exploitation, and cruelties of their fully human superiors.

What has become evident as scholars have traced out these histories is that the centrality of the category of the human to the forms of recognition and misrecognition that have structured and oriented our social lives and political institutions did not come from an embarrassing failure by some people to be expansive and broadminded in their understanding. Rather, the use and misuse of the category of the human has been bound up with the strategic imperatives central to the projects of nation-building and imperial expansion (Alcoff 2005; Butler 2010, 2006, 2004; Collins 1998; Hirschmann 2013; Stoler 1995, 2010; Weheliye 2014). This bitter history has meant that even though the putative openness of the category of the human has made it available for appropriation by a range of excluded groups in their struggles for equality (Bunch and Frost 2000; Nussbaum 2000; Reilly 2007), the category *as a category* has also come under critical scrutiny. For in order for the category to be at once open ("everyone") and exclusionary ("not everyone"), the qualities that define the human must be held as criteria against which claimants to ethical and political recognition must be measured (Brown 2004; Esposito 2012). It is not the case that one simply is human. Rather, one must be deemed so. To be human, then, is not a state of being or an attribute but rather an aspiration, an attribution, and an achievement.

The critical analyses that conceive of the human as a politically potent achievement turn on the insight that the category depends upon, and cannot be thought without, its comparative and differentiating functions (Esposito 2012). If the human is an implicitly comparative category, against what is it measured? If it entails a movement of differentiation, from what? And toward what? Three threads of argument that emerge from many and sophisticated efforts to interrogate these movements of comparison and differentiation focus on humans as an exceptional species, humans as bearers of rights, and humans as agents in a field of action. I will sketch each thread in turn.

The first thread of argument concerns the long-standing question of what differentiates humans as a species from other animal species. And what scholars have found is that in the Western tradition of thought at the very least, humans' self-understanding as special, especially deserving, or exceptional rests on their somehow exceeding whatever qualities or capacities they share with animals. Elizabeth Grosz observes that Western philosophy and science have "attributed to man a power that animals lack"—the power of reason and speech, of intelligent response, or the moral capacity to experience shame (2011: 12). The imperative to differentiate humans via comparison is so strong that no sooner are bees, crows, orcas, or chimpanzees found capable of such "human" arts as dance, creative thinking, cultural formation, or war than these criteria for distinguishing the human are replaced by other characteristics or activities deemed exceptionally human. Giorgio Agamben (2003) suggests that the inexhaustible search for criteria by which to distinguish humans from animals indicates that what is sought and valued in that search are not the capacities or characteristics themselves but rather the gap between animal and human that the capacities or characteristics are thought to evidence. As a consequence, Agamben says, we can see that "man is the animal that must recognize itself as human to be human" (2003: 26). Indeed, Agamben proposes that the practices of compulsive differentiation that produce the gap and thereby the human should be known collectively as "the anthropological machine" (2003).

The imperative to find a definitive difference between humans and animals—and the inability satisfactorily to do so—is not just a ghostly relic of past cultural and political efforts to demarcate and differentiate humans in ways that underwrite and justify colonial expansion, slavery, and eugenic management of populations (Esposito 2012; Foucault 1978, 2003, 2009; Schiebinger 1993). The project of spotting the difference continues today in contemporary science—and there is no remedy to be found there either. As the genomes of humans and other creatures have been mapped and studied, scientists have learned that because of our evolutionary history, many genes among bees, mice, chimpanzees, and humans are shared (Bejerano et al. 2004; Sharan et al. 2005). The conservation of genes and patterns of protein usage across species suggests that no one gene or set of genes definitively marks human difference. The difference is a matter of degrees, perhaps, rather than of kind (Grosz 2011). Similarly, recent archeological findings suggest that modern humans, that is, Homo sapiens, did not arise through a singular differentiation from their historical forebears but rather from mingling, living, and interbreeding with them. In other words, there is no human as a pure species because modern humans are a variegated hybrid of homo species, an amalgamation of Neanderthal, Denisovan, and other Homo variants (Aiello 2010; Callaway 2014; Lordkipanidze et al. 2013).

One might claim that the search for what differentiates humans from animals seems like old-fashioned nonsense because really, in everyday life, it is quite obvious which creature is a human and which is not. But what is at issue in the historical and theoretical analyses rehearsed above is precisely that what counts as human has not been obvious—and remains questionable for many nonnormative or marginal populations, such as homosexuals and transsexuals, comatose medical patients, people with physical or mental disabilities, racial and ethnic minorities, immigrants and refugees, prison inmates . . . the list goes on (Agamben 1998; Baynton 2001; Butler 2004; Esposito 2012; Hammonds and Herzig 2008). The critiques that trace the difficulty of differentiating the "truly" human from its nonhuman others suggest that rather than denoting a creature per se, the category of the human designates a constellation of rights, duties, and prerogatives that attach to those who recognize one another as worthy of carrying them.

So, the second thread of arguments examining how the human is defined via comparison and differentiation focuses on the designation of those rights, duties, and prerogatives. From this perspective, the human is conceived not as an essence of existence but as a particular form of coexistence. As Stanley Cavell (1979) might say, in this view, to be a human is a matter of acknowledgment within a relationship rather than a matter of knowledge. The category "human" here is an index of a moral dignity that requires consideration and that obliges interactions characterized by respect.

But to think of the category of the human as a kind of index of moral dignity and rights does little to save it from disintegration under critical scrutiny. The reason is that if the dignity and rights indexed do not attach to something essentially distinctive in humans, then the quality of regard, dignity, and respect to which the term refers can also attach to nonhuman creatures—to primates, to dogs and cats, to cows, sheep, and pigs, to animals generally (Cavell et al. 2008; Singer 2009; Sunstein and Nussbaum 2005). If the quality of regard, dignity, and respect commandeered by the category of the human do not attach necessarily and narrowly to those creatures we have called human, then the category suffers a kind of boundary failure. As Cary Wolfe notes, after scholars and activists repudiated assumptions that inclusion in the set of "the deserving" be restricted to select groups based on features such as gender, class, ethnicity, race, or national origin, "it was but one short step . . . to insist that species too should be set aside" (2003: xii). In this move, the expansion that abrogates the anthropocentric "rights like us" perspective renders the nominative "human" obsolete: the rights, obligations, and duties qua rights, obligations, and duties no longer require the modifier "human" to organize them or to be meaningful and persuasive. The collapse of the human as a qualitative category also leads to its collapse as a sign of moral exception.

The third set of arguments challenging the ways the human is defined via comparison and differentiation is closely related to the ethical revaluation of nonhuman creatures outlined above. Implicit in Habermas's plea to retain the human as a moral category—to uphold human exceptionalism—is not simply a desire to see the human as an exceptional repository of distinctive rights and obligations but also a desire to preserve the idea that humans are exceptional actors or agents (2003). What is at issue is the idea of humans as a very particular kind of agent, one who is distinct from the field in which he or she acts, and one whose actions are contained or delimited by his or her intention and self-conscious deliberation. But for decades, scholars have dismantled the notion that humans are agents of this sort. Pointing to the play and structure of language, to the mutual shaping of reason and the passions, to the embedding of economic norms, social expectations, and cultural conventions in the intimate depths of desire, psyche, and flesh, and to the reverberation of individual and collective actions through social and material space and time, scholars have discredited the notion that humans are self-mastering, that their actions are characterized by deliberation and autonomy, that their intention is realized in and contains the action that is its effect.

Critics of the notion of the human have mobilized these challenges to propose that there are many more agents afoot in the world than human exceptionalism has allowed. The idea here is that if human actions are conditioned by manifold social, material, institutional, and corporeal factors, if those factors contribute to, amplify, redirect, inspire, undercut, and make possible or impossible human activities, then those factors should also be considered as possessing what Diana Coole (2005, 2013) calls "agentic capacities." The argument is not that we should grant to water resources, social networks, ecologies, germs, monetary systems, or digital technologies an agency they did not previously exercise. Rather, it is that we should recognize and bring within the ambit of our theoretical work the fact that they have always been efficacious in their activities in ways that conventionally have been captured under the rubric of agency (Bryant 2011; Connolly 2011; Coole 2013; Delanda 1997; Haraway 1991; Harman 2011; Hayles 1999; Latour 2007; Morton 2013a, 2013b; Stiegler 1998). Indeed, considering the ways that humans and nonhuman creatures and objects only together exert effects to transform the terms and possibilities of our coexistence, Jane Bennett proposes that we relinquish the idea of *individual* agents and think instead in terms of "a heterogeneous assemblage" that produces effects through its various and changing interrelations (2010: 23).

This kind of intervention resolves what has been thought as merely the context or field of human action into heterogeneous elements of a broad network of enabling and constraining factors that must be ignored or abjured if humans are to conceive of themselves as self-sovereign subjects. Whatever node, spark, energy, or force we might have thought as the distinctively human capacity to bend intention into action becomes, in such analyses, yet another myth or illusion, a narrow hypostasizing of many causes and forces that serves to conjure the human as the origin of its actions and as the master of an action's realization and effects. Critics have zeroed in on the presumption that humans-as-agents transcend "the bonds of materiality and embodiment altogether" (Wolfe 2010: xv). And they counter this presumption by positing "a certain nonhuman agency as the condition of possibility of human agency" (Bennett 2010: 98). They contend that what makes humans effective is not their differentiation from their embodied existence or from the social and material contexts of their actions but precisely the opposite condition, to wit, "the irreducible imbrication of human/nonhuman or natural/social processes" (Coole 2013: 454).

Together, these three threads of argument call on us to think carefully about what we are saying when we talk of humans as exceptions to the animal rules, when we invoke humans as a group or the human as a moral and political category. The critiques do not amount to a denial that there are human creatures in and on this world. Nor do they add up to a claim that there is nothing whatsoever that is distinctive about human creatures in the ways that they live, love, and die. As Sharon Krause rightfully notes, to acknowledge the distributed forms of efficacy by which material environments, social institutions, creatures, ecologies, and technologies condition human action is not necessarily to deprive humans of any capacity to act or to absolve them of responsibility for what they have done or continue to do (2011). Rather, these critical challenges to the concept of the human constitute an effort to show that, as Cary Wolfe puts it, "the 'human' . . . is not now, and never was, itself" (2003: xiii). The human as we have tended to think it is built around a profound refusal to admit or to acknowledge some of the muddy and messy conditions of existence that humans share with all living creatures. The challenges, then, call on us to rethink and animate differently the category's founding elisions and conditions, to elucidate and then to refigure the contours and texture of our self-understanding so that the work the concept of the human does in our thinking does not carry and elaborate what is noxious in its previous assumptions (Wolfe 2010: xvi).

In an analysis that participates in this chastening of the human subject via an elaboration of the agency particular to objects of various shapes and sizes, Timothy Morton (2013a) claims not only that humans are far from being the most important protagonists in the emerging disasters of global climate change and environmental degradation but also that specifically "nonhuman beings are responsible for the next moment of human history and thinking" (201). Indeed, in a particularly pessimistic coda, he suggests that the evident impotence of humans-as-agents combined with the enormity and momentum of the looming catastrophes means that our predicament is best captured by the primeval relation of a mortal to an unpredictable and impetuous god (201). But it seems to me that to cringe, pray, hope for the best, and prepare for the worst is to cede too much, to throw in the proverbial towel.

For what is at issue here is not simply whether and the extent to which we can transform our modes of living in such a manner as to ameliorate rather than compound the problems with climate and environment. Nor is it whether we can leverage social and scientific technologies to limit the effects of the damage we have evidently wrought. We also need to consider how to mitigate the ways that the accretion of political and economic injustices creates specific forms of vulnerability to disaster for particular regional, national, and subnational populations. This need is compelling, since many of the deleterious effects of global warming and environmental degradation are structured and riven by human, that is, social and political, categories and relations. As Jesse Ribot (2014) remarks, as important as it is to attend to the geological effects of climate change, such a focus can also divert attention from "the grounded social causes of precarity that expose and sensitize people to hazard" (668). In other words, Ribot claims, what makes climate change and environmental degradation disastrous in many communities is to a large degree institutionalized obstacles to broad-based social and political self-determination. To meet the challenge of the disasters is to have to come to terms not only with what humans do to the world they inhabit but also with what they do to and with one another. The inequalities and the differential forms of vulnerability to which they give rise are social and political products to be problematized and countered rather than taken for granted, naturalized, and accepted as an inevitable state of affairs (Ribot 2014).

As we face the burgeoning and unpredictable effects of global climate change and environmental degradation, as well as the forms of conflict and devastation that accompany them, the various challenges to the concept of the human demand that we consider and consider again what kinds of actions we can and should take. If environmental and climate catastrophes demand a response from a specifically human subject, as Chakrabarty contends, the breadth and depth of criticism aimed at long-standing conceptions of such a subject suggests that we tread carefully as we imagine the contours and characteristics of this critically important political actor. What we need in the place of the fantasy of human exceptionalism is a different figure of the human, one that does not succumb to the conceits of old but also does not conceptually dissolve humans as identifiable agents and thereby absolve them of the crises that mark the Anthropocene.

The counter-concept of the human that is implicit in the critiques outlined above is one of a creature who is an embodied and thoughtful animal as well as a technological aficionado, a creature embedded in and composed by the social and material contexts of its existence, an agent whose actions are dependent on and conditioned by manifold networks of ecological, institutional, social, and symbolic relations. In this book, I endeavor to make this implicit refiguration explicit. I turn to the life sciences to sketch the basis for this refiguration in part because so much of what drove the old project of the human was a revolt against embodiment, against the animality, the organismic, the materiality of human creaturely existence. Indeed, what I find so exciting and intriguing in turning to the life sciences to think about this refiguration is the convergence of insights in social theory with grand and profound shifts in scientists' understandings of what humans are. Like social theorists, scientists increasingly confirm that there are complex interactions and interchanges between biological and social processes that muddle any distinction we might want to make between body and environment. And as Celia Roberts notes in her study of hormones, the research that traces these binary-busting processes is being used to develop models that provide a radical reconfiguration of who and what we are (2007: xv). Nikolas Rose proposes that the convergence between critical theoretical, social scientific, and emerging life science understandings of the human demand that we revisit questions such as "how we should live as humans, why we should live as humans, of what we owe to ourselves and others, of what we can know, what we should do, what we can hope for" (2013: 23). In many respects, this book is the beginning of an answer to just such an invitation.

"Wait! Science?"

Of course, the idea that one might draw on the life sciences to help develop the basis for a new theory of the human could cause many a theorist to issue a spluttering, snorting cough in surprise. The discomfiting ambivalence toward science that fuels such an aspirated eruption has roughly two dimensions.

The first dimension elaborates the concern that, today as in other historical periods, the sciences as a form of knowledge are inextricably bound up with the forms of political economy and population management that exacerbate and perpetuate inequality and injustice. Even if scientists do not actively or knowingly collude with nefarious political actors, they are immersed in, absorb, and articulate their questions and findings in the terms that structure and animate historically specific norms and political ideologies (Code 2006; Franklin 2000; Harding 2004; Hubbard 1990; Montoya 2007; Shapin and Schaffer 1989). For instance, Nikolas Rose (2007) argues that biotechnological developments prompt individuals to conceive of themselves and to associate with others in terms of biological characteristics such as disease, injury, bodily modification, or the travails of reproduction. The conglomeration of investment, research, and service institutions dedicated to providing assistance in such matters encourages individuals to become adept at self-management. And in a fashion that Wendy Brown claims is the signature of a neoliberal political economic formation, this push to self-management turns individuals away from broad-based collective political action in favor of issue-specific group lobbying efforts that meld together expertise, scientific technologies, and corporatist forms of fundraising (2005). Similarly, as Duana Fullwiley (2015), Jonathan Inda (2014), Kim Tallbear (2013), Dorothy Roberts (2011), and Anne Fausto-Sterling (2005) argue, even as scientists might disavow racism and anticipate that genomic sciences could undermine any claim that racial inequalities have a biological basis and explanation, the forces of the pharmacological-biotechnological market combine with linguistic habits and histories of racial inequalities to (re)consolidate the worn and dangerous idea that race is bound up with genetics.

At the heart of this concern, then, is the realization that every form of knowledge is shaped by reigning ideological imperatives and that no form of knowledge—the sciences included—can be insulated or isolated from the cultural, moral, and political tensions that define and texture collective life (Haraway 1991; Lewontin and Levins 2007). The worry is that the use of findings in the life sciences as a resource for reconceptualizing "who and what we are" could result in a stealth importation of racial, gender, or other ideological biases into work that is oriented precisely around the historicization of and critical challenge to them. Even if, as Nikolas Rose (1998) suggests, we credit scientists with working within a framework that is "normed by truth and characterized by a philosophy of veridicality" (161), such a framework cannot account for the ways that dominant political norms and assumptions work their way into science unreflectively and unintentionally. If such assumptions inflect science in spite of the efforts of numerous scholars and activists to raise awareness of how they infiltrate the practices and institutions through which scientific research proceeds, then it would seem naïve, foolhardy, or even somehow complicit to make a resource of science research without at the same time constantly calling it into question.

The second dimension of the ambivalence toward science is elaborated around the concern that a turn to the life sciences for the purposes of a theoretical refiguration of the human could serve to bolster the already pervasive aggrandizement of the sciences as the only legitimate or worthwhile form of knowledge. Scholars in the fields of science and technology studies have done great work tracing the social, political, and economic dynamics through which scientific practices are constituted as authoritative and as a superior form of knowledge production (Åsberg and Birke 2010; Code 2006; Hird 2004; Kirby and Wilson 2011; Longino 2002). The authority attributed to scientific modes of knowledge production translates into substantial corporate and foundation financing for research and institutional stature for its academic practitioners. Such financing and stature are so proportionally great compared with the humanities and social sciences that the scholars in the latter fields find themselves sidelined or marginalized in their academic institutions—their work little understood, poorly resourced, and generally undervalued by administrators.

Maurizio Meloni notes that these complex dynamics of authority, stature, and resource provision can lead to the perception that the biological sciences, with their emphasis on empirical findings, aim to "colonize the social" (2014: 733). And indeed, Meloni points out, quantitative social scientists have turned to the life sciences with the idea that it could provide "a foundational vocabulary" (734) for the study of social and political life, with the idea, that is, that biology could "authenticate" (739) the concepts and terms that more qualitatively oriented scholars might use to develop their analyses and arguments. For scholars who do qualitative research, such an imperialist ambition is objectionable in (at least) two ways. First, it demeans the rigor of the methodologies and the generativity of the insights that turn on critical reflection about individual, collective, and historical forms of self-understanding. Second, it presumes—absurdly—that those vastly complex and manifold forms of self-understanding that we call culture could be reduced to and explained in the empirical terms of biological functioning.

These two general concerns about the inevitable complicity of scientific findings with political imperatives and the widespread mission creep of scientific modes of knowledge production are extremely important. Nonetheless, I do not think they justify a refusal to engage with the intriguing developments in contemporary science. The reasons are threefold.

First, there is the remarkable convergence of contemporary science with some of the best and most critical forms of theoretical work on the insight that there is no "pure" biosubstrate, with the consequence that "bio-" is not something to which an explanation of any phenomenon can be reduced. The theoretical challenges to the concept of the human that highlight what William Connolly (2013b) describes as "our manifold entanglements with nonhuman processes, both within the body and outside humanity" (401) are mirrored in scientists' fairly trenchant reconsideration of the relationship between living organisms and the worlds they inhabit. Rabinow and Caduff (2006) remark that whereas geneticists historically have tended to focus on the genes within cells as the key determinant of creaturely growth and behavior, a growing number now also look for explanation beyond the enclosure of the cellular nuclei housing the genes—and beyond the confines of the body itself. Instead, they attend to "the complex interactions between cells, systems of cells, multicellular organisms, populations of organisms and their environment" (330). Confirming this broader sense of what is included in the domain of biology, George Slavich and Steven Cole (2013) observe that the complex interdependencies evidenced through these interactions have undermined scientists' "longstanding belief... that we are relatively stable biological entities... [who] live in a dynamic social environment" (343; see also Robinson et al. 2005; Robinson et al. 2008). It turns out that rather than conceiving of cells as hermetic units that protect their contents from the vagaries of the world, "our molecular 'selves' are far more fluid and permeable to social-environmental influence than we have generally appreciated" (Cole 2009; Slavich and Cole 2013: 343).

In this emerging view, the embodied human does not simply move in a field of action but absorbs manifold substances from its habitat and responds perceptually, biochemically, and viscerally to the threats and promises that the social and material world present (Landecker 2011; Landecker and Panofsky 2013; Lock 2013; Niewöhner 2011). Scientists conceive of this traffic into the body and its cells as nutrients, gases, chemicals, and toxins that transit the skin, the mucous membranes, and the cellular membranes of the body through breathing, eating, and absorption (Guthman and Mansfield 2013; Mansfield 2012a). Perhaps more strikingly, they also conceive of it as "nurture, culture ... geography, experience, and history," as Fausto-Sterling puts it (2008: 683), in other words, the structured, experienced, imagined, or anticipated social encounters and relations that form the texture of daily life (Kuzawa and Sweet 2009; Slavich and Cole 2013; Thayer and Kuzawa 2011). Indeed, Slavich and Cole point out that together, the material and the social and cultural habitats in which a living body develops, grows, and lives is a considerably more significant determinant of which genes are used in the body and how they are used than what we could call "in-born" genetic factors (333). Julie Guthman and Becky Mansfield (2013) rightly note that these findings, which indicate that the processes of development and growth in organisms are shaped by myriad environmental factors, are "paradigm-shifting" (491)—and are all the more remarkable for their consonance with shifts in social and political theory.

Elaborating some of the implications of such studies, Guthman and Mansfield (2013) observe that "there is nothing about the body that forms a solid boundary—or threshold—between it and the external environment" (497). In these kinds of analyses, the body is conceived as porous or permeable in a way that belies the sense either that the "self" is distinct from the body or that the body is distinct from the social and material environment (Landecker 2011; Lock 2013). Rather than containing some portion that is "given" in biology and some portion that is "made" through culture and environment, the human body is increasingly conceived as hybrid, as constituted through "active processes that are simultaneously political and biophysical" (Guthman and Mansfield 2013: 490). So pervasive is this emerging sensibility that scholars are now trying out various ways to capture this simultaneity, offering such terms as *transcorporeality* and *viscous porosity* to indicate the traffic between body and environment (Alaimo 2010; Tuana 2008), *nature-cultures* or *naturecultures* to indicate the thoroughly mixed or hybrid domains we inhabit (Haraway 2008; Latour 1993), and *biocultural* (Davis and Morris 2007; Fausto-Sterling 2008; Goodman and Leatherman 1998; Saldanha 2009) and *socionatural* (Guthman and Mansfield 2013) to point to the combination of forces and factors that inextricably together compose human life.

For theorists as well as contemporary scientists, then, humans are constituted through a matrix of biological and cultural processes that shape one another over various time scales in such a way that neither one nor the other can be conceived as distinct. As in social and critical theory, so in the sciences this sense of the reciprocal shaping of body and environment has reached a point that Susan Oyama (2000a) and Evelyn Fox Keller (2010) caution against talking as if the environment and the body *interact* to shape one another: the conjunction "and" that binds the body and the environment grammatically in such a formulation places the two as a priori distinct phenomena that (then) come together. However, Keller notes that "the image of separable ingredients continues to exert a surprisingly strong hold on our imagination, even long after we have learned better," which is to say that there is nevertheless a persistent tendency among scientists and social and political theorists to think and work as if there is "a space between nature and nurture" (2010: 30). And in an effort to counter such a tendency, Karen Barad has suggested that the concept of "intra-action," or action within a field, might better capture the processes by which the biological and the social reciprocally work on one another (2007).

The broad outlines of the emerging scientific consensus about the porosity or permeability of the biological body are so coincident with the criticisms of the disavowal of embodiment, corporeality, materiality, and interdependence made by theorists challenging the category of the human that some form of mutual engagement would promise to be productive. For just as those who have critically reconsidered the concept of the human carry forward presumptions and habits of thinking that have not yet been transformed, so too do scientists. For us to engage with one another's work through this moment of convergence seems to present an opportunity for each kind of critique to be thrown back on its assumptions, to discover new modes or perspectives for thinking, to become bewildered only to perceive a novel pattern or unexpected set of connections.

The second and related reason not to refuse to engage with developments in contemporary science concerns how theorists might take up the research findings. The engagement with science represented in this book does not amount to a call for theorists to become scientists or to become the "so-what" brigade tasked with elaborating the policy, regulatory, or administrative ramifications of particular experimental findings. The idea is not at all for theorists to learn the science so as to formulate a plan of action, as if we should adopt a purely instrumentalist orientation toward this form of knowledge with the presumption that "knowing will lead to knowing what to do" (Weigman 2010: 84). What is at stake in the engagement proposed is not an epistemological issue in which particular claims can be adduced true or false, verified or disproven. What is at stake is instead something akin to an ontological issue that addresses itself not to questions of truth or falsity but to our orientation to our world, to the patterns and connections we trace between phenomena, to the background framework within which the meaning of particular statements and claims can be assessed (White 2000; Zerilli 2005). The engagement with and creative appropriation of the general outlines of the human to be found both in the theoretical critiques of the concept of the human and in contemporary science can elucidate in new ways the terrain of our social and political lives. In doing so, they can "[bring] forth the ethical character of every act" and "[expose] the possibility and necessity of decision" (Fynsk 2004: 76).

Contemporary scientists are finding increasing evidence to support the claim that culture, symbolic forms of communication, and imaginative anticipation shape the ways that bodies compose and recompose themselves over time (Cole 2009; Kuzawa and Sweet 2009; McEwen 2012; Slavich and Cole 2013). As in social theory, so in the life sciences, culture, environment, embodiment, and self are suddenly uncertain terms whose confusion and disorientation reverberate to similarly unsettle cognate concepts. The engagement with science that is partially undertaken here and that is offered for other theorists' imaginative, creative, and critical appropriation is a project of theoretical experimentation, an effort to push and rework those uncertain terms, to try to effect a gestalt shift that will generate new possibilities for thinking about politics.

A third reason for theorists to engage with the life sciences as we reimagine and refigure human being is that theorists' critical acumen is needed. Our dexterity with different frameworks for analyzing the ways that language, space, identity, norms, and cultural practice shape people's experiences of their material and social habitats must be brought to bear as scientific researchers continue to develop their own insights. The convergent shift in thinking about the human shows us that the social, economic, material, and ecological environments we inhabit shape us in our growth, development, interactions, and capacities in ways that are just as profound as the norms and disciplinary practices that invite, produce, and constrain our individual and collective identities. To broaden the terrain of our analysis to include this traffic will enable us to think more deeply about embodied subjectivity and the incorporation of norms and to do so in ways that might enhance the kinds of analysis of "intelligibility" and "framing" with which we are currently more familiar (Butler 1993, 2010). Similarly, by taking into substantive account the ways that the materiality of the environment as well as our responses to social interactions—real, symbolic, and imagined—become embedded in the body's composition and habitual function, we can enrich our theories of affect to include the ways that accumulated experience can entrain and constrain as well as enable future forms of mutual engagement (Leys 2011; Papoulias and Callard 2010).

Further, as this material enhances our understanding of biopolitics, it can also illuminate new fields for democratic politics. For instance, Guthman and Mansfield (2013) propose that the scientific findings indicating that the environment enters and transforms the body point to a need to think in terms not only of "the molecularization of life" through which biotechnological advances are sold as solutions to our problems but also of "the environmentalization of the molecule" (491). In other words, pushing against the constraints posed by extant theories of biopolitics, they explain that the chemical molecules that enter and transform the substance of our bodies and selves do their work not just in the laboratory or the clinic but also in the environment—the environment here conceived as chemical, microbial, and fungal pollutants, health systems, patterns of nutrition, access to water, programs of waste disposal, and so forth. Accordingly, we should treat "the mutable, biological body as being constituted not only through intentional intervention and management but also through interactions with the wider environment" (500). One possible consequence of such understandings of the constituting activity of the environment could be the portrayal of individuals as responsible for their exposures (Landecker 2011; Mansfield 2012b)—a recapitulation of neoliberal biopolitics. But such an entailment is not a necessary one. Because these environments are shared, held in common through place and/or culture, the effects they have on cognition, health, judgment, social interactions, and habitual patterns of behavior—all of which are bound up with the capacity to participate in politics—can be conceived as collective problems with political and possibly democratic modes of resolution. And in the same way that the modes by which the social and material environment enters and transforms the body "are unlikely to be race-, class-, or gender-neutral" (Guthman and Mansfield 2013: 500), so too will researchers likely vehiculate the norms, assumptions, and forms of representation that constitute this crazy unjust world in which we live. Heeding Clare Hemmings's (2005) cautionary reminder that culture is not a seamless entity, we have to bring to bear on our appropriation and elaboration of this life science research our critical awareness of the ways that norms and presumptions about race, gender, sexuality, and disability, the directives of nationalist ideology, and the imperatives of the economy together shape both the social and political worlds we inhabit and the ways we understand and write about ourselves.

A More Personal Beginning

I came to this project as a scholar who had spent some time making sense of the materialist dimensions of Thomas Hobbes's philosophy and theory of politics. In one of those wonderful refractions of thought that come from the bizarre conjunctions of graduate course work, I read Monique Wittig's rants about the materiality of language and embodiment at the same time I revisited Hobbes's *Leviathan* and *De Corpore*. It became clear to me that Hobbes was a trenchant materialist, yet his political theory was consistently read as if he were an adherent of Cartesian rationalism. So my doctoral dissertation and my first book set out to explore what would happen to Hobbes's ethical and political thinking if we were to grant him his account of the subject as wholly embodied rather than as somehow split between body and mind, or body and soul (Frost 2008).

After that project, I wanted to continue to think about how a materialist understanding of the self might reshape our understanding of politics—and I hoped to do such thinking in a contemporary vein rather than continue working with Hobbes. However, I realized that whereas Hobbes provided a fairly well-elaborated account of the body from its minute bits to its gross form, I did not have a good grasp of current accounts of either the molecular or the systemic processes through which living bodies persist in their existence. So I took some courses in biology during a sabbatical and then secured a generous fellowship from the Andrew W. Mellon Foundation that funded an additional length of study, with the result in the end that I took various life science courses full-time for eighteen months.

I worked through organic chemistry, molecular genetics, and the biology of perception, through biochemistry, cell signaling, endocrinology, and the metabolism of brain function, and to courses on primatology and environmental toxicology. I sat through some fascinating lectures, read lots and lots of textbooks, talked with my professors, and engaged in longer conversations with some of those same professors as my colleagues.

I would like to be able to say that I was a gracious and composed student. But I wasn't. As a professional academic and a parent coordinating childcare and the general continuance of daily life, I didn't have enough time to study in the way I remembered being able to study, which was frustrating for the nerd in me. The language and concepts initially were so alien that the readings took me forever. Without a background or commonsense understanding of what was at issue, I did not know at first how to select, distinguish between, or remember the relevant or important pieces of information. I was aghast at how stressful test-taking was, particularly because I did not anticipate how the questions on the tests would take forms different from those in the humanities courses I generally teach. It was bracing, exciting, and energizing. I was humbled, humiliated, shocked, and wonderstruck.

Eventually, I became more conversant in the concepts, more familiar with the abbreviations and acronyms. In the middle of the training the world picture started to congeal, and I was able to perceive, play with, and anticipate patterns in the material I encountered. And this led to a different kind of difficulty—which was how to relate my new-found knowledge to political and cultural theory. At times, new information or insights from the scientific studies would throw me back on my habits of thinking, my theoretical training, my critical proclivities, and leave me bewildered about what I know.

And of course, that bewilderment became a running theme in my intellectual life after my stint taking courses was over. I had to figure out what to do with it all. I had not simply absorbed an enormous amount of information; I had acclimated to a new conceptual vocabulary, inhabited a different mode of being in and looking at the world. But what in the world would or could I do with it?

Over the past decades, the body has been beautifully rendered as a cultural artifact, as a product of norms, culture, discipline, and power, as ground, matrix, or extension of socially mandated and carefully elaborated forms of subjectivity. To say "You are a body" as a theorist is to invoke all that work. It seems a different thing, a more complicated thing theoretically, to say "You are a living body (you are alive)." With some notable exceptions, I think theorists are more comfortable with general or theoretically abstracted notions of embodiment as manifest in the notions of corporeity, vitality, liveliness, and generative becoming than we are with notions of living bodies persisting with and because of biological functions and processes (Ahmed 2008; Connolly 1999, 2002, 2011, 2013a; Davis 2009; Haraway 2008; Hird 2009a; Kirby 1997; van der Tuin 2008). To describe someone as a living body, as alive, seems at once to be absurdly obvious and also theoretically suspect—as if the statement of something so clearly incontrovertible must carry another agenda, that is, the aim to reduce someone to a biological substrate (the implication of course being that liveliness is something to which humans could be reduced, that humans are more than our biology . . . a presumption that in turn rests on the conviction that biology and culture are fundamentally, substantively, qualitatively distinct).

In reflecting on her efforts to persuade scientists to think beyond the gene—to incorporate the complex life cycle and organisms' social and material environments in the study of development and evolution—Susan

Oyama wryly notes, "Theorists are annoyed when they are told what they have 'always known.' Yet there is a difference between knowing in a parenthetical 'of course it's important' way ... and incorporating the knowledge into models and explanations, research and theory" (2000a: 200–201). Oyama's identification of the "parenthetical 'of course'" is an allegory for what theorists know about being alive and what we are willing to integrate into our critical and theoretical work. Even though we can accept that we are embodied creatures and even though we often advocate for an understanding of the ways that the meanings we attribute to bodies accrete to become the very experience of the flesh, there is also a broad tendency to presume a mutual exteriority between the human and biological self. Observing such a tendency, Georges Canguilhem notes that "man sometimes marvels at the living" and yet is also sometimes "scandalized at being himself a living being" (2008: xix). This sense of scandal, of impropriety, of disapprobation and indignation reveals the force of this presumption of mutual exteriority, suggesting that it is fairly fundamental to the ways we have made sense of ourselves and the world.

This book is a step toward facing that scandal, an effort to begin thinking theoretically about humans as alive, as living bodies, as biocultural creatures.

Biocultural Creatures: Structure, Puzzles, and Concepts

In the text that follows, I draw on different dimensions of the life sciences to think through and recast the conceptual terms and the figures according to which we understand what humans are. Very broadly speaking, my aim in this book is to exploit the convergence of scientific findings with contemporary theory to elaborate a granular, substantively detailed ontological warrant for thinking the human as a theoretical and political category in the context of posthumanism. In the course of doing so, I provide some language, conceptual terms, and theoretical movements that might prove helpful in challenging the theoretical habits that sometimes lodge in our thinking and that make it difficult fully to occupy the idea that humans are biocultural creatures. But even as the book lays out some of the groundwork for reconceptualizing humans as biocultural creatures, it is important to note at the outset that humans as a distinct species and as a particular kind of subject do not actually appear much in the pages that follow. Explaining why will help clarify what I am trying to do.

This book does not contribute to the literature that displaces the human by denaturalizing the categories through which conventionally it is understood. In playing conceptually with molecules and proteins, it does not broach a nonanthropocentric starting point as a lesson-in-perspective for the habitually and chronically anthropologically minded, as is so provocatively done by Myra Hird (2009b) in her investigation of bacteria or Michael Marder (2013) in his reflections on plant life. Rather than chastening the human, this book takes as an instructive point of departure the posthumanist critiques of the category of the human, filling out the conceptual hunches at the heart of those critiques. It is because this book is an effort to figure humans in a way that does not exclude materiality, "objectness," animality, or embeddedness in habitats that it starts with energy and atoms and works up through the scales of molecules, cells, proteins, to gross organisms . . . and ends by merely gesturing to humans.

Another way to elucidate this particular strategy is by noting an observation made by Pheng Cheah (1996) in one of his early critical engagements with theories that focus on the materiality or corporeity of the subject. With an impressive depth of analysis of arguments offered by Judith Butler and Elizabeth Grosz, Cheah notes that considerations of the matter of corporeity or of encultured embodiment often turn on or operate with an anthropocentric starting point. He suggests that a more thoroughgoing transformation of our understanding of the materiality of embodiment might turn on a reconfiguration not of the relation between culture and the body but of the distinction between form and matter (134). In a sense, the trajectory of this book is the result of a what-if, as in, what if I were to start with quantum physics, which fuzzes the very distinction between form and matter. The very first chapter, on carbon, begins with the formulation that matter or substance is an effect of energy that is constrained in the ways it can relate to itself. The effects and manifestations of energy-constrained become a running theme through the book, drawing attention to the dynamic processes and activities that constitute living embodiment at different spatial and temporal scales.

Further, and related, I want to note that in elaborating on the energetics of matter and living processes, I do not intend thereby to disavow the forms of representation, the norms, and the myriad social and political practices that organize spaces and inflect modes of living to an extent that they become a part of who and what people are. Rather, holding on to the insights of contemporary theory, I hope to fill out a dimension of human subjectivity that is frequently acknowledged as complex but not often given much bona fide texture (Frost 2014). I am trying to give substance and theoretical form to Susan Oyama's insight that the "interdependence of organism and environment" entails that they "define the relevant aspects of, and can affect, each other" (2000a: 3); I am trying to make good on Elizabeth Grosz's claim that it is because we are alive that we can participate in "social, political, and personal life" (2004: 1–2). In short, I hope to provide concrete details and a set of theoretical figures of movement that may enrich our understanding of the processes through which humans in particular live in and become acculturated to and through their worlds.

In experimentally rebuilding our imagination of the human, I have written this book so that the chapters follow a cumulative logic. Because of this logic, the book's insights are more readily to be conjured through a straight read rather than selective picking. Each chapter is oriented around both a puzzle raised by my encounter with the science and a key concept that contributes to a refiguration of the human. In most cases, the putative resolution to the puzzle accomplished by each chapter is bound up in the puzzle that organizes and motivates the next. I give fairly robust overviews of the chapters here so that they can serve as a resource to orient readers as they work through the narrative details of the text.

Chapter 1 stages an encounter between on the one hand contemporary theoretical fascination with materialization, corporealization, animation, and vitality and on the other what we know ("of course") about the nature of matter in our nontheoretical, daily-engaged-with-the-world kinds of ways. The chapter starts from the puzzle that while matter or materiality is currently conceived as an important starting point for thinking about embodiment and politics, in physics and chemistry, "matter" is conceived as energy. What exactly does it mean to say that matter is "really" energy? What would we have to adjust in our theories of materiality, materialism, corporeity, or embodiment if we were to conceive of matter in terms of energy? Drawing on insights from quantum physics and organic chemistry, this chapter proposes that energy takes form as matter through its constrained self-relation. What we know and experience as matter is energy under a particular form of constraint. This chapter uses this insight to outline the reasons that carbon is the kind of matter that forms the basis of life as we know it.

Chapter 2 builds on the first chapter's explanation of the ways that energy is constrained in its self-relation to reconsider the nature of boundaries and identity. It does so by meditating on the way that the constraints on energy that make it take form as matter also compel matter's arrangement into porous cell membranes. The puzzle here concerns that membrane porosity: if a cell membrane is constitutively porous, which is to say that it is composed in a manner that enables a rapid and constant flow of biochemicals into and out of cells, then what kind of boundary is it? This chapter proposes that what makes the puzzle puzzling is that we tend to think about boundaries that distinguish one side from another in terms of substances. What makes a cell or a living creature a discrete cell or living creature if its "inside" is constituted through a continual chemical exchange with its "outside"? This chapter suggests that we should eschew the tendency to think of boundaries in terms of a topographical demarcation of substances. It proposes instead that we think about the boundaries effected by the composition of cell membranes as creating conditions for distinct kinds of activities. The permeability of a cell membrane facilitates a constant and continuous influx and efflux of chemicals into and out of a cell, which enables a distinctive and distinguishing kind of cellular activity to take place. In this case, the "inside" and the "outside" established by cell membranes mark not different substances but different zones of activity. This chapter proposes that this theoretical formulation enables us to begin to imagine how creatures might be embedded in and constituted by their environs or habitats yet also identifiably distinct from them.

Chapter 3 broaches anxieties about biological reductionism that often motivate a cautionary orientation toward biology or lurk beneath a reluctance to draw on aspects of the biological sciences for thinking about cultural and political phenomena. This chapter refigures the nature of biological matter and biological agency by tracing the composition and activity of the proteins that facilitate and regulate the traffic of substances into and out of cells through those porous membranes. In elucidating how genes are used in the composition and activity of those proteins, the chapter engages the puzzle of how a biochemical process involving molecules can unfold in precise, specific, responsive, and reliably repeated ways without there being some kind of agent who makes it happen. It explains that each step in a biochemical process depends on both a prior action and the availability or arrival *en scène* of biochemical molecules from elsewhere in the cell, from outside the cell, and possibly from beyond the boundaries of the organism. The prior biochemical action and the availability of specific biochemical ingredients constitute both the enabling and the limiting conditions that make a specific step in a biochemical activity possible. The conceptual formulation that captures this chapter's insight is that because of these enabling and limiting conditions, biological processes have direction—but direction without intention. The chapter proposes that this formulation enables us to account for the precision and directedness of biological activity without that activity being reducible to anything at all. The formulation also makes it impossible to identify any part of the body that is "purely" biological: the entirety of a living organism is biocultural.

Chapter 4 draws on the insights of the previous chapters to bring into sharper focus the relationship between an organism and its habitat. Revisiting the idea that biological processes and their constituent biochemical activities are possible because of the traffic of biochemicals across the boundaries effected by porous cell membranes, this chapter traces what oxygen does in a living body. Pushing against the tendency to see ingestion as compositional (in the sense of adding stuff to the body), the chapter argues that the activity of oxygen requires that we conceive of bodies as "energy in transition," as a system of processes that mobilize and take advantage of the ways that energy subsists and transforms under constraints. The puzzle engaged by this chapter concerns the vital importance of oxygen: if every step in a biochemical process is a critical condition for the next step, and if a failure in any step could mean a failure in the biochemical process, why is the aspiration and use of oxygen so very critical for survival? The chapter suggests that while every step in every biochemical process in a cell is the condition of the next, the presence of oxygen in a cell is an index of the whole organism's exposure to and utter dependence on its engagement with its habitat. The theoretical point of the chapter is that we can only properly appreciate the micro-level biochemical processes that constitute living if we remember that they occur in whole organisms who engage with and are dependent on their habitats.

Chapter 5 begins with the recognition that the insights of chapter 4 take us extraordinarily close to the position that a living organism is not much more than—and is possibly reducible to—its interactions with its habitat. This chapter thus reconsiders whether we can think of living creatures as discrete organisms rather than as effects of the environmental relationships and dynamics in which they live. Put slightly differently, the chapter engages the puzzle of how to conceive of the "it-ness" or fleshy reality of bodies in a context in which we conceive of bodies in the terms of energy in transition and interchange with habitat. If living bodies are porous and are constituted through their continuous engagement with their habitats, how can we conceive of them as distinct from their environments rather than merely reducible to them? The chapter engages this puzzle by exploring the temporality of the relationship between organisms and their habitats. The chapter explains how the permeability of germ cell membranes entails that a creature's response to a habitat in one generation can shape the development and growth of offspring in subsequent generations via epigenetic markers. The chapter proposes that this anticipatory, cross-generational carrying-forward of creatures' engagement with and response to their habitats indicates that bodies are noncontemporaneous with their habitats. It is this noncontemporaneity that makes organisms exceed the determinations of their habitats even as they cannot be thought apart from those habitats.

Together, these chapters spell out some of the concepts, movements, and figures through which we can imagine humans as biocultural creatures. It is just a beginning. I hope that the experiment in thinking that the book as a whole represents provides resources for other thinkers to do creative and politically generative work.

ONE

Carbon

*In which we learn that energy takes form as matter
through its constrained self-relation*

Right from the very start of my ventures into biology, I found myself thrown back on my operative assumptions and having to rethink the concepts I had thought I would use. In the researches on embodiment and subjectivity that I intended to bolster and elaborate by studying biology, scholars explore what it might mean for our self-understanding and for politics if we figure matter, material things, and fleshy bodies not just as objects but as agents, as possessing some kind of agent-like force or capacity. According to such work, matter should be conceived not merely as an inert passive phenomenon that is buffeted by and that vehiculates the (human) powers external. Rather, it has a force or effectivity immanent within it. Manuel Delanda (1997), for instance, explains that "matter is much more variable and creative than we have ever imagined," for "even the humblest forms of matter and energy have the potential for self organization" (16). Much of this generative creativity or self-organization evinced by matter comes from "interactions between parts" through which "properties of the combination as a whole . . . are more than the sum of its individual parts," properties that are thereby described as "emergent" (17). The impetus behind this shift in thinking is to displace the anthropos as the privileged locus of agency, for "if matter itself is lively," as Jane Bennett speculates, "then not only is the difference between subjects and objects minimized, but the status of the shared materiality of all things is elevated" (2010: 13). To disconcert ourselves by perceiving and tracing the vibrancy of matter could enable us to see the world and our place in it differently.

But as I began taking the organic chemistry courses with which all biology students generally begin, I was compelled to acknowledge theoretically some things about matter that actually I had already known but had neglected to incorporate into my materialist schema. What I needed to acknowledge is captured well in Timothy Morton's recent observation that whereas in everyday life we conceive of matter as "the 'out-of-which-it's-built' of an object.... When you study it directly, it ceases to be matter" (2013b: 62). The first point of acknowledgment was that matter is not matter-in-general but matter-more-specifically. Matter is composed of elements whose varied composition constitutes the conditions of possibility for material things' being, persistence, transformation, and effectivity in the world. In other words, in working with notions of materiality, we have to conceive of matter not as a relatively undifferentiated mass of substance but rather as a broad array of atomic elements each of which is composed quite differently and specifically, as elements whose very specificity has a profound effect on how each behaves.

Moreover, and this was the second thing, what makes the atomic elements different in their specific ways is that they are composed of variously charged subatomic bits and pieces. Which is to say that when you get right down to the details of it, the atomic elements that compose matter are not really "stuff" at all but rather conglomerations of energy—a fact that complicates considerably the question of what we are to understand matter to be, and what it might mean to position oneself theoretically as a materialist. If researchers in a variety of fields are beginning to reconsider the significance of materiality, if they are taking matter as a bona fide starting point for their investigations, what might such a troublesome atoms-are-energy hiccup portend? Disconcertingly, I found myself having to challenge and transform my understanding of materiality at the very start of my effort to equip myself so as to be able to persuade more scholars to take it seriously.

It turns out, however, that if we give ourselves over to the insight that the atomic elements that make up matter are composed of energy, we do not lose the ability to talk about materiality. To the contrary, to attend to energetic basis of matter in the way that I outline in this chapter enables us to understand how matter is differentially composed and how different kinds of matter behave and participate in processes not just in random ways but in marvelously, astonishingly specific, complex, and directed ways. Indeed, to think about the energies out of which matter is composed enables us to appreciate the reasons that carbon is the predominant matter of the matter that is the body, that is, that life as we know it is a carbon-based form of life.

In this chapter, I will explain how different forms of energy constrain each other in ways that end up being generative of the elements we know as matter. The particular figuration I want to advance is that energy takes form as substance or matter through its constrained self-relation. This is a very compact formulation, to be sure. But once its meaning is elaborated, it is also very productive. For the constraints on energy that make possible its substantialization as matter are the very constraints that are generative of life processes. To conceive of matter—organic matter and organic processes—in terms of the constraints on the energy that give it form enables us to wrest productive insights from many aspects of contemporary biology, insights that animate the rest of this book.

The Stuff of Matter

One of the interesting things about matter is its uncertain ontological status as substance. In spite of several generations of advances in physics and chemistry that take for granted the energetic qualities of matter, cultural and political theorists—and perhaps the lay public more generally—tend to hold on to a substantialist framework for thinking about matter. Within this framework, matter is a substance with extension, density, and duration. And it is composed of smaller bits or particles that congeal, organize, or become differentiated so as to create the substances and solid objects we experience in our everyday lives. Such a tendency to conceive of matter as substance is, perhaps, part of the legacy of modern atomism, according to which the characteristic movements of gross physical objects are read back into their constituent parts. As with many things from modern philosophy that retain their hold on our forms of understanding, atomism and substantialism accommodate the seeming solidity of objects we encounter: a rock, a hand, an apple. Such objects feel solid—solid and hard, solid and soft, solid and crisp.

Of course, we have also heard tell of electrons, protons, and neutrons, with their electrical charges and subatomic and quantum component parts that might eventually reveal secrets about the origin of the universe. Yet, because it is difficult to think about energy as substance or material, we tend to ignore those findings and gravitate in our common theoretical parlance toward a substance ontology: it just makes better and easier sense of the heft, the weight, the pressure and resistance that rocks, hands, and apples exert as we grasp them. To consider what it might mean to think about matter in terms of energy is a fascinating challenge not only to the ideas with which we often work but also to our experience of ourselves in the world.

The simplest way to begin is to imagine holding two magnets together, not on their sucked-together attracting sides but on their pushing-away repulsing sides. There is a squidgy bouncy quality to the repulsion. The magnets slide around the gap between them; or perhaps it feels like a gap holds them apart. What keeps the magnets apart is magnetic force, which is a kind of energy. Even though magnetic force is a kind of energy, it feels solid when under pressure. Indeed, in our working scenario, we only feel this seeming solidity when we use our hands to hold or press the magnets toward one another. The polarity of the magnets' energies compels the magnets to move apart, and it is when they are compelled by our pushing hands to be in proximity that we experience a sense of solidity or substance. Using the magnets as an imaginative reference point, we should think about matter in this way, as forms of energy whose interrelations are constrained or delimited in such a way as to create what we know as substance.

To think about matter as energy that feels substantial because its movements or shifts are curbed or circumscribed is not to say that the solidity of matter is "not real," as if up to this point we have been mistaken or deluded. Rather, it is to say that the solidity or substantiality of the matter we encounter daily is an effect of the constrained flow and interrelation of energy. Although admittedly, if we were to dig theoretically down to the quantum level, as Karen Barad (2007, 2012; Barad and Kleinman 2012) invites us to do, then we would look at the world through a framework in which it is possible that all that is solid could melt into air (a horrible cooptation of Marx's lovely phrase). For as Barad points out, at the quantum scale, the different dimensions of energy split and rejoin at the same fraction of an instant, so that each quanta both exists and does not exist at the same time: "a dynamic play of in/determinacy" (2012: 214). In thinking through the lens of quantum field theory, then, the solidity of matter might well be a delusion. But since the indeterminacy, instability, and unpredictability of quantum behavior has posed difficulties for scientists who might aspire to integrate it into other branches of scientific research, like biology, I will not dwell in this book at that level of analysis.[1]

So, to formulate a definition that can serve as a reference point for our discussion: energy takes form as matter through its constrained self-relation. What we know or experience as matter is energy whose differentiation produces highly constrained forms of self-relation. Those highly constrained forms of energetic self-relation are the conditions for the generation of various forms of extension, density, endurance, and dimension, some of which are beyond human perception but some of which we humans experience as heavy, light, staid, evanescent, solid, fluid, airy, opaque, or transparent. Some of those forms are hot, some cold; some soothing and some are explosive. And among those forms is a very special one: life.

Our effort to conceive of matter as variegated formations produced through the constrained interrelations of different kinds of energy will be helped along if we refigure or revise our imaginative conception of what an atom looks like. For a long while, atoms have been portrayed via a solar system model: just as our solar system has a sun around which circle various planets (and protoplanets—poor Pluto), so an atom was understood to have a center or nucleus around which orbited little particles known as electrons. It turns out, however, that such a figuration does not do justice to the form and movements of the energy that composes atoms: it makes atoms seem like particles that are made out of tinier particles that themselves are made out of even tinier ones.

Instead, if we consider that the different "pieces" of an atom are forms of energy, then we can think about the shapes, dimensions, and behavior of atoms in terms of the interactions that take place between those different forms of energy (just as in the case of the imagined magnets). As I will show, the movement of the energies that compose electrons, neutrons, and protons—and, yes, that then compose atoms—is not an easy circulation, a lazy circuit, or a gently rhythmic flow. Rather, the movements are constrained and contained, such that they constitute fields or domains that agitate tensely. It is these tense fields of energetic agitation that form the components of the atoms that we can perceive when they coagulate into objects and sensory organs. If we hold in our imagination the thought of those magnets—drawn together yet held apart, repelled from one another but prevented from dispersing, moving and shifting within these dual constraints—we can appreciate better the kinds of energy that compose an atom and the ways that those energies interact to create what we know as matter.

Outline of an Atom

At the center of an atom is the nucleus. The nucleus is a zone or field of energy composed of neutrons and protons. And while neutrons and protons are not as important for my overall discussion as the electrons that circulate around them, there are a couple of features that are helpful to understand. Thus, an outline rather than a lot of details.

Neutrons are particle-like forms of energy composed of a variety of quarks—the quarks are "various" because they are themselves different forms of energy. Neutrons have a mass and no charge. One of the jobs that neutrons perform is the provision of the force (called "the strong force") that holds the protons together in the nucleus. Protons are also tiny particle-like forms of energy composed of a (different) variety of quarks. They have mass, and they also possess a positive charge. In fact, it is because protons have a positive charge that neutrons are so handy. Just as the positive ends of two magnets repel one another, so the positive charges of protons force their separation. Indeed, the conglomeration of protons in an atomic nucleus would violently repel one another but for the fact that they are held together by the strong force of the neutrons. The nucleus, then, is not a still or quiet field. Rather, it is characterized by a trembling tension generated by the neutron's strong force that holds the protons together despite the repulsion that would otherwise push them apart.[2] The constraints on energy in relation to itself gives form to the centers of atoms.

There are a couple of other interesting things to know about atomic nuclei.

First, exactly what kind of substance or element an atom is depends solely on the number of protons in its nucleus. The elements are not differentiated because of some enduring substantive essence—carbon, for instance, is not carbon because it is and remains over time "carbonish." We have to dispel any residual Aristotelian notion of "species" here. Rather, the elements are differentiated—different from one another—because they differ in the number of protons they possess. If we start with one proton, we get hydrogen. If we imaginatively add one more proton to the nucleus, then we get helium. If we serially add one more proton, then one more proton, and so forth, we would, or could in principle, traverse the entire table of elements, with each addition creating an atom of the next kind. We could also, in our imagination, do the reverse and trip down through the table of elements from huge heavy atoms to the smaller ones—all the way back down to hydrogen. The fact that proton count makes each element what it is effects a peculiar and somewhat disorienting de-essentialization of the different elemental forms of matter. The distinctive and what we often think of as identifying characteristics of each element—oxygen, chlorine, silver, or lead, for instance—depend on the number of positively charged protons in an atom's nucleus and the kinds of (energetic) interactions among protons and electrons that that positively charged population makes both possible and impossible. The constraints posed by the ways that energy can relate to itself are what constitute the differences among different kinds of matter.

To appreciate that proton population is what defines an element is to make sense, albeit an anachronistic one, of the completely disreputable but centuries-old and widely practiced art of alchemy, and particularly the effort to chemically transform common metals into gold: if one could adjust the number of protons in the right way, one could create a veritable pile of treasure![3] Historically, this sort of alchemy was doomed to failure because practitioners tried to use chemistry (which is concerned with electrons surrounding the nucleus) rather than specifically nuclear chemistry (which is concerned with the particles inhabiting the nucleus). But even today, this sort of alchemy is close to impossible—or at least the attempt is not casually recommended. It is close to impossible in part because of the prohibitively enormous amounts of energy required to shove just one additional proton (let alone a whole slew of them) into a nucleus. It is also better not done because the reactions precipitated inside the (tense) nucleus by such an addition are incredibly energetically powerful, difficult to contain, and oftentimes explosively dangerous. Such changes in composition of an atomic nucleus are called nuclear reactions; the energy that is released and that can subsequently disrupt the composition of other atomic nuclei is the phenomenon called radioactivity (Atkins and Jones 2010: 707).

A second interesting feature of atomic nuclei is the fact that, generally speaking, at least in the smaller atoms, protons and neutrons are numerically matched. That is to say, except in the case of hydrogen, which has one proton and no neutron, if there are x number of protons, there are x number of neutrons, and vice versa. But sometimes there can be an imbalance between them. This can happen when the addition of energetic particles to or the subtraction of energetic particles from the nucleus of an atom changes the form of energy that composes the quarks. When the quarks that compose protons and neutrons change, protons can change into neutrons or neutrons into protons (Atkins and Jones 2010: 708). When this kind of transformation, which is called nuclear transmutation, results in only an increase in the number of neutrons, the atom in question is considered "heavy." However, its heaviness does not change the kind of element that the atom is: as noted, "element-ness" is the province of protons. However, an atom's heaviness can render it somewhat unstable and subject to spontaneous nuclear transformation. And when nuclear transmutation results in a change in the number of protons, a different kind of atom has been created.

Interestingly, this kind of nuclear transmutation is actually a fairly common phenomenon in nature and indeed is the key to the technique of carbon dating. When the sun's energy enters the earth's atmosphere, it collides with the nuclei of various atoms in the upper atmosphere. Those collisions release neutrons (among other things). Those rogue neutrons strike into the nuclei of nitrogen atoms, adding a neutron to the neutron population and in the process forcing the ejection of a proton. As a result, the nitrogen atom (which generally has seven protons and seven neutrons) is transformed into a carbon atom (with six protons and eight neutrons). This carbon atom is considered heavy because generally carbon atoms have six protons and six neutrons. Now, because heavy carbon has six protons—and also six electrons—it participates in just the same way as regular carbon in the biochemical processes that make living plants and creatures live. As a consequence, heavy carbon comes to be a part of living plants and creatures—it is integrated into all carbon-based life forms—in proportion to the amount of heavy carbon in the atmosphere. Over time, various energetic shifts around and within such a heavy atom will evict the "extra" neutrons, thereby turning the heavy carbon into a regular carbon (with six protons and six neutrons). In fact, the decay of heavy carbon into normal carbon occurs at a steady rate.[4] When an organism dies and its carbon composition is no longer in equilibrium with its environment (no eating, no equilibrium!), it will gradually have less and less heavy carbon than a similar living organism. So, because heavy carbon is created at a fairly constant rate in earth's atmosphere and because living organisms have a fairly constant ratio of heavy to normal carbon in them, we can measure the age of a dead organism by how much its heavy to normal carbon ratio differs from that of a living organism (Atkins and Jones 2010: 719).

The last part of my outline of an atom concerns electrons. The next section will elaborate particular features of electrons and their interactions with one another in more detail. But before that detail, we need a more general picture. So: electrons are the third major constituent of atoms. They are generally numerically matched with protons and have a distinct and polar opposite charge, designated as a negative charge. And whereas protons and neutrons are localizable in the nucleus, electrons exist as movement around the nucleus. Note, here, that I say "they exist as movement" and not "they are moving"; the latter would imply a particle. Electrons are particle-like, but they do not have any mass; indeed it is because they have no mass that we are even more hard-pressed to figure them in other-than-particle terms. They are perhaps best imagined as a shifting flow of negative energy. As shifting flows of negative energy, they do not occupy specific or fixed locations. Of course, this sounds a little bizarre: they exist, but nowhere in particular.

Perhaps Barad's evocative account of electrons can help us grasp the meaning of such a seeming paradox. Barad explains that for quantum field theory, the electron is a "self-interaction," in which "the electron's self-energy takes the form of an electron exchanging a virtual photon (the quantum of the electromagnetic field) with itself" (Barad and Kleinman: 212). What she means is that rather than simply being a stable instance or movement of negative energy, an electron shifts, splits, and rejoins itself "not sequentially" but rather in the same moment or instant (Barad 2012: 80). This self-interaction makes the electron determinate and indeterminate at the same time, a condition she indicates via a slash: in/determinacy.

It this in/determinacy that makes an electron both exist somewhere but also not at the same time.

Another way to begin to imagine the meaning of such a claim—although one that does not do justice to the quantum level of analysis preferred by Barad—is by means of an admittedly poor but nevertheless usefully everyday analogy. Take your coffee cup, half full. Swirl the coffee. As you swirl, the amount of coffee in the cup does not change, but at one moment there is a "high tide" of coffee on one side of the cup while there is also a "low tide" of coffee on the other; split seconds later, the high tide and low tide are in successively different areas of the cup, always somewhat opposite one another and always on the move. Get a really good swirl going, and you can see the bottom edge of the cup while the coffee travels the cup's perimeter. Note that while the coffee is swirling around and around, you can't specify definitively that the coffee is "right there," for instance, "by the cup handle." It is somewhere in the cup, moving, skipping over itself, sloshing around . . . and therefore is nowhere in particular even as it will likely pass the cup handle periodically in its swirling movement. If you could manage to take a snapshot of the swirling coffee just as it reached the area by the cup handle, then you could indeed say: it was here at this point in time. However, in the image it would no longer have that momentum or movement (and so wouldn't be the swirling coffee that it is), and it would have moved on in its swirling path in the cup and therefore wouldn't be there any more, despite its capture in representational form at the moment of the snapshot. This is akin to what electrons are like. They swirl or spin in a fairly defined region around a concentration of protons in a nucleus. Accordingly, while they are "there" in that region somewhere, their particular location in that spatial field is a matter of probability, except in retrospect.

In thinking about the matter that composes life, in considering the constraints on and the shifts in energy that drive biological processes, the parts of the atom that we are interested in are the protons and the electrons. It is the protons and their associated electrons that give atoms their capacity to interact with one another to create substances and to generate reactions between them. The numerical balance between protons and electrons—and the forces of attraction and repulsion that they exercise on one another with their positive and negative charges—are the key factor in chemical reactions. Those chemical reactions enable atoms to attach to one another to build large complex phenomena. Equally significant, those reactions also enable atoms to transform their interrelations so that in their joined form they change shape or energy formation, bend or move, or become differently constrained in an astonishing variety of ways to form the substances we know as solids, liquids, and gases, rocks, plants, and creatures.

In order to understand how the shifting, containment, and redistribution of proton and electron energies is critical to the liveliness of living matter, we need to examine further the disposition of electrons around a proton-filled atomic nucleus. The energy that composes electrons relates to itself and to protons in ways that generate the tensions, intense frissons, and pathways to abatement that together produce atoms with distinctive characteristics, molecules with a particular form and reactivity, and the grosser forms of matter whose activity is the basis of life processes in organisms. To gain insight into the tensions between electrons that at once pull atoms together and hold them apart is critical to grasping how the composition of organismic flesh creates not a solid impermeable boundary but rather one with gaps, fissures, crevices, crevasses—and sometimes outright holes—all of which make possible and sometimes facilitate the traffic of molecules in and out of cells, in and out of bodies. Although it will take some work to get there, to make sense of the disposition of electrons is to have the basis for understanding and articulating how all creatures are biocultural creatures.

Electrons in Particular

Electrons, which have a negative charge, are held close to the nucleus of an atom by the attraction exercised by the positively charged protons housed there. However, despite the fact that the proton-filled nucleus exerts an attractive force on the electrons, the electrons do not all simply zoom straight into the center of the atom to occupy the nucleus.[5] The reason they do not collapse into the positively charged nucleus concerns the fact that electrons are waves of negatively charged energy. Since electrons repel one another with their negative charges, they together prevent one another from getting close to the positively charged nucleus to which they are attracted. They arrange and shift themselves around the attractive proton-filled nucleus in patterns that ensure that they are somewhat close to the nucleus yet interfere minimally with one another. If they were to go closer to the nucleus, they would become frenzied in their energetic movement because of their mutual proximity. So, as Lederman and Hill (2011) explain, generally electrons persist least frenziedly in what is sometimes called a ground state, where there is a balance between the pull of the protons and the repulsion between the electrons (296–97). Here we see again: constraints on the relation of energy to itself give form to the perimeters of atoms.

The patterns by which electrons arrange themselves around a nucleus are fairly standard across different atoms, and they have two features. First, there are layers moving from the middle of the atom outward, which are called shells (visualize concentric ripples created by dropping a pebble into a pond). Second, within each shell is a set number of areas or spatial fields in which a set number of electrons may spin. These spatial fields are called orbitals. Now, given the foregoing discussion, it might seem that the repulsion exercised between electrons in proximity would entail that no two electrons could persist in the same spatial field or orbital. However, electrons do not all spin in the same, uniform direction. They spin up. Or they spin down. In fact, we could say that they spin left or right, or in or out of the plane of this printed page. The point is that electrons spin in one of two distinct and opposite directions that scientists have labeled up and down. Because of their spin, two electrons can occupy the same orbital if they spin in opposite directions. So, each orbital in an atom may contain up to, but no more than, two electrons with opposite spin.[6]

So far, then, we have electrons conceived as waves of negatively charged energy drawn to the positively charged nucleus, constrained from reaching that nucleus via mutual repulsion, spinning in fields known as orbitals. The pull, the repulsion, the tension in balance: it all contributes to the composition and interaction of atoms. To get one step closer to those interactions, we need to examine the shells mentioned briefly earlier.

Because there can only be two electrons per orbital, and because the arrangement of orbitals around a proton-filled nucleus is limited by their resistance to one another, an increase in the number of protons and electrons in an atom entails an accommodating increase in the number of layers or shells that form concentrically around a nucleus. The more protons and electrons, the more shells. The distribution of electrons in those shells is standard across elements—whether we are talking about relatively small atoms like lithium, oxygen, or carbon (which have just two shells) or relatively big atoms like tin or lead (which have four or five shells) (Lederman and Hill 2011: 166). The first and innermost shell has one spherical orbital, which, because it is just one orbital, may contain only up to two oppositely spinning electrons. The second shell, which is further removed from the proton-rich nucleus than the innermost shell, has four orbitals. Given the limit of two oppositely spinning electrons per orbital, then, the second shell may have up to eight electrons. The orbitals in the second shell are shaped and arranged in an optimum pattern for the mutual noninterference of the electrons. One of the orbitals is spherical (roughly surrounding the first shell); the others are shaped like ring doughnuts if mapped in three dimensions.[7] Of the three doughnut-ish orbitals, one goes left-right (the x axis), one goes up-down (the y axis), and one goes in-out of the plane (the z axis) (Anslyn and Dougherty 2006: 4–5; Lederman and Hill 2011: 166–67). There are third and fourth shells too, which also have their distinctive and standard sets of orbitals, but these are for elements whose atoms are not often involved in the organic chemical reactions we are concerned with as we learn about biological processes. So I am going to ignore them.

Now, what is important to understand about the shells is that when each is full—either the first shell with its maximum two electrons in the one orbital or the second shell with its maximum eight electrons distributed through four orbitals—the balance between the attraction of the protons and the mutual noninterference of the electrons is so perfect that the atom is inert: nothing disturbs it. Atoms of elements that are composed of complete shells do not react with others. They are chemically inactive.

However, when the shells are not completely full, either the first shell with just one electron or the second shell with between one and seven electrons, then the balance between the attraction of the protons and the mutual noninterference of the electrons is imperfect. There is an imbalance between the strength of the attraction evinced by the positively charged protons and the effects that the negatively charged electrons have on one another. Such an imbalance is what makes atoms interact or react with one another: the energetic conditions that constitute the imbalance draw atoms toward one another and hold them variously closely in ways that at least temporarily resolve the imbalance. When atoms that are imbalanced interact in ways that resolve that imbalance, they are said to have engaged in a chemical reaction: chemical reactions are the shifting and sharing of electrons between internally unbalanced atoms.

To explore what it means to conceive of the energy imbalances between protons and electrons in this way, consider this: the more protons there are in the nucleus, the stronger the attractive force they exert on the atom's electrons. Yet the more protons there are, the more electrons there will be too, which means that some electrons will spin in the "outer" or second shell rather than in the "inner" or first shell. The further from the nucleus the electrons are located, the smaller is the effect of the proton attraction on them.

So, if we imagine an atom with three protons and therefore three electrons, the third electron is in the second shell: it is furthest removed from the attractive force of the protons and thus held more loosely than the two electrons in the first shell. If we also imagine an atom with nine protons and therefore nine electrons, two electrons would be in the first shell, and then seven would be in the second. These seven electrons would be held closer to their nucleus than the lone electron in the three-proton atom because of the attractive force of nine protons rather than of three holding those electrons close. Indeed, the force of the proton attraction would be so great that the nine-proton atom could pull away the lone second shell electron from the three-proton atom. Such a theft would fill up its second electron shell to its maximum and stable eight. It would also render the smaller atom relatively more stable because its electron pattern would be akin to ground state.

But that is not the whole story.

The nine-proton atom, with its stolen electron, would have a numerical imbalance between the protons and the electrons: it would have an extra electron. This extra electron makes the atom as a whole have a negative charge. Similarly, the three-proton atom, missing its lost electron, would also have a numerical imbalance between protons and electrons: it would have more protons than electrons. This surfeit of protons makes the atom as a whole have a positive charge. The two atoms, then, one with a net negative charge and the other with a net positive charge, would be drawn and held together by electrostatic attraction. Indeed, this is exactly what happens to create salt. Sodium, with three protons, and chlorine, with nine protons, engage in just this electron switch to form sodium chloride or common table salt.

The conditions under which electrons persist—the attraction and draw of protons combined with the mutual repulsion that compels their particular arrangement around an atomic nucleus—are the conditions that make possible the interaction between atoms, interactions that are the basis for many complex substances with many complex forms. It is when atoms lend, share, or steal the electrons in one another's outer shells that we say they have undergone a chemical reaction: chemistry is just this. And the force that holds the atoms together as a result of a reaction is called a bond, a chemical bond. Because chemical bonds bind atoms together into what we experience as matter, we need to understand them more fully in order to appreciate how energy takes form as matter through its constrained self-relation. We also need to understand them because the formation and transformation of chemical bonds is the basis of organismic life processes.

Bonds

Anslyn and Dougherty (2006) claim that, in its most general sense, a chemical bond is an overlap of atomic orbitals (4). However, not all bonds are alike, for the manner in which the electrons are shifted or shared defines and shapes the kind of bond between two atoms. Three kinds of bonds concern us here: covalent, ionic, and hydrogen bonds.

Covalent bonds are those in which the electrons in question are *shared* between two atoms. Here, it is not the case that one atom deprives the other of one or some of its electrons. Rather, the atoms blend some of their not-yet-full orbitals so that when the shared electrons spin, each atom can be said to have completed its outer electron shell. In the same way that a shared driveway between two houses turns what is technically half a driveway for each house into practically a whole driveway for each, so in covalent bonds, the electrons in the outer shells of the atoms are integrated into one another's orbitals such that, for all intents and purposes, each involved atom has in its outer shell eight electrons.

The atoms conjoined by covalent bonds are together called a molecule. A molecule can be composed of anywhere between two, fifteen, thirty-seven, or more than five thousand atoms. The electrons involved in a covalent bond spin in orbitals that stretch between the bonding atoms, circulating around and between the nuclei of both. The orbitals that contain these shared electrons are called molecular orbitals. And because the electrons shared in the molecular orbitals complete the electron shell for each involved atom, the atoms joined in a covalent bond are in a lower energy state—they are figuratively less tense and more restful. As a result, a covalent bond holds the atoms together strongly: a covalent bond is quite difficult to break.

One of the things that is interesting about covalent bonds is that they can be single, double, or even (though rarely) triple. The reason this variety of covalent bonds is interesting is that each kind contributes differently to the shape of the resultant molecule. And, importantly, as in the poking of round pegs in round holes, the stacking of bowls, or the insertion of a key into a keyhole, the shape of a molecule affects whether and how it can interact with other molecules. In other words, the shapes of molecules are in part an effect of the length of the bonds joining the atoms. The attractive force of the protons of one atom on the electrons of another, the mutual repulsion of the protons in the distinct atomic nuclei, and the mutual repulsion of the electrons in the conjoined atoms together affect how long or short the bonds will be and the extent to which the bonding atoms will bend at different angles to accommodate the various pulls and tensions. The number of orbitals involved in a covalent bond shapes the angles at which the conjoined atoms lie vis-à-vis one another.

A single bond occurs when a covalent bond forms involving just two electrons. This is called a sigma bond (as in the Greek letter σ). Sigma bonds are notable for the fact that the atoms involved in them can twist around: just as your arms can rotate in their sockets to point your hands forward, up, sideways, back, or down, so sigma bonds enable atoms in a molecule to rotate around the bond in different directions.

Sometimes a covalent bond forms in which two electrons from two orbitals in one atom enter into a bond with two electrons in two orbitals from another. This is the case with oxygen. An oxygen atom has six electrons in its second shell. If we imagine distributing those six electrons to the four orbitals in that second shell, we get two full orbitals and two half-filled orbitals—which means that the atom needs two more electrons to get to its magically stable eight. To do this, two oxygen atoms can blend their sets of two half-filled orbitals to create two molecular orbitals each with two electrons. In such a case, the two blended or molecular orbitals, which are two covalent bonds, stretch in a roughly parallel fashion between the involved atoms. This double bond is called a pi bond (as in the Greek letter π). What is notable about pi bonds is that their parallel structure does not allow for rotation. This means that the atoms involved in them tend to lie flat or in a plane vis-à-vis one another.

What is especially interesting about double bonds is that the concentration of electron energy in the vicinity of the bonds is serially redistributed through many or all of the covalent bonds in the entire molecule. This redistribution is called resonance. One of the fun things one learns to do in organic chemistry is to trace the possible patterns of resonance that are created in molecules. One traces those possible patterns because as the bulge in the electron charge shifts around the molecule (remember the coffee in the coffee cup), it leaves other areas of the molecule with a slightly more positive charge. The shifting of electron charge through resonance renders some atoms of a molecule slightly more negative and some atoms slightly more positive. This in turn makes it possible—or makes it more or less likely—for the molecule in question to enter into particular kinds of reactions with other atoms or molecules, a process that thereby creates larger or different molecules.

Triple covalent bonds are very unstable—think of such a situation as too much of a squeeze in the sense of compelling electron orbitals to bunch closely together in parallel formation on one side of the adjacent atoms. So rather than being a common feature of molecules, they are often transitional, which is to say that they form and quickly unform as a set of molecules goes through a series of reactions in order to get from A to Z. Accordingly, they are often associated with changes or transformations in shape.

Ionic bonds are very different, as was illustrated in the example of salt or sodium chloride. In the formation of an ionic bond, an atom with an especially proton-rich nucleus steals an electron from an atom with a noticeably proton-poor nucleus. In this exchange, the proton-rich atom performs as an electron acceptor, and the proton poor atom performs as an electron donor. After the exchange, both atoms become what are known as ions—one a positively charged ion (called a cation—pronounced cat-ion) and the other a negatively charged ion (called an anion—pronounced an-ion). What is important to remember here is that while the electron shells of both anions and cations are stable, each such atom, as a whole, is energetic because of its charge. As a consequence, each is readily drawn to oppositely charged ions and held close to them by mutual attraction. The holding together of anions and cations via such electrostatic attraction is what constitutes an ionic bond (Jones and Fleming 2010: 5; Lederman and Hill 2011: 174). Because an ionic bond is formed through electrostatic attraction rather than through the sharing of electrons, the bond is not as strong as a covalent bond. Indeed, if an ion with a particularly strong charge were to come into contact with two atoms conjoined via an ionic bond, the ionic bond between the conjoined two would break, while a new ionic bond would form between the greater attractor and the ion attracted to it. This is what happens when water dissolves salt.

Finally, *hydrogen bonds* are a weak but biologically important form of bond. Hydrogen bonds generally occur not between free-floating atoms but rather as a result of a particular situation in an already formed molecule. A covalent bond can sometimes be unbalanced in the way that the bonding electrons are shared—not unbalanced enough to qualify as the theft involved in an ionic bond but an unequal distribution of the electrons' negative energy. This kind of covalent bond is known as a polar covalent bond. Polar covalent bonds occur when the large number of protons in one atom's nucleus shifts the molecule's net electron charge toward that portion of the molecule. This shift leaves the proton-rich atom in the bond with slightly more negative electron charge around it and the other atom in the bond with slightly less negative electron charge around it—or, we could say, the positive charge at the core of the latter atom is more effective at affecting other atoms.

When a polar covalent bond involves a hydrogen atom, which is especially small with just one proton and one electron, the shifting of the bulk of its electron energy toward the atom with bigger proton count makes its own proton significantly less buffered by negative electron charge. As a result, the hydrogen portion of the molecule exudes positivity. Thus, when this molecule with an extroverted hydrogen atom comes close to another molecule that also has a polar covalent bond, the more positive hydrogen exerts an attractive force on the relatively negative region of the other molecule, an attraction that holds the two molecules fairly close together (Atkins and Jones 2010: 178–79; Nelson and Cox 2008: 43–46). To be sure, they are held together fairly loosely. One would not say that a hydrogen bond conjoins atoms or molecules. But even though hydrogen bonds individually are not very strong, they often occur in aggregate—and in aggregate they are strong enough for their effects to endure over time. They can bend a big molecule into a particular shape or hold two large molecules together. Indeed, the proximity between atoms and molecules that hydrogen bonds effect is crucial to many aspects of biology. As I shall show in chapter 3, hydrogen bonds are critical to the folding of molecules into proteins that participate in manifold and very specific kinds of biological activity.

Why Carbon?

Having seen how the constraints on the relation of energy to itself is constitutive of atoms, chemical reactions, and molecules, we are now positioned to appreciate why carbon is the form of energy-as-matter that serves as the basis for life itself. The primary elements involved in the molecular chemical reactions at the center of life processes are those that have just one or two electron shells radiating from their atomic nuclei. Because these atoms are fairly small, their nuclei are fairly stable and their proton count fairly consistent. Such stability and consistency is crucial for chemical reactions that take place again and again over time. Indeed, in many respects, such constancy and regularity are what make repetition possible.

There are six atomic elements that form the building blocks of the molecules that together compose life: hydrogen, carbon, nitrogen, oxygen, phosphorous, and sulfur. And we say that life on earth is a carbon-based form of life because carbon atoms form the backbone or the scaffolding that structures the molecules that together make life possible.

So, why carbon, as opposed to any of the other building blocks? If we keep in mind the ways that the constraints on the self-relation of energy constitute the conditions for the formation of atoms and molecules, there are several reasons we can surmise for carbon being the matter that matters so much. And in each case, it is the constraints on what carbon can do that enable it to do so many things.

One. It turns out that carbon is special because it is a rather middling atom. As noted, as atomic elements increase the number of protons in their nuclei, the number of electrons in their shells also increases. Those atoms with just one or two electrons in their second or outer shell are extremely likely to lose or donate those electrons to another atom. Such atoms are very reactive electron donors. Conversely, those atoms with six or seven electrons in their second or outer shell are extremely likely to grab or steal an electron from another atom. They are very reactive electron acceptors. Strikingly, carbon is right in the middle of that lineup: with six protons altogether, it has two electrons in its first shell and four distributed through the orbitals in its second shell. Possessed of four and in need of four in its second shell, carbon can either lend or attract electrons, depending on the other kinds of atoms surrounding it (Nelson and Cox 2008: 496). In other words, carbon is flexible in the way it participates in bonds and therefore is flexible in its range of possible bonding partners.

Two. Another ramification of carbon's middling proton count is that it is difficult to make an ion out of a carbon atom. The relationship between its protons and electrons is such that the nucleus of a carbon atom is not strong enough to pull an electron from another atom. At the same time, though, it is strong enough to prevent another atom from easily making off with one of its own electrons. In other words, it is tricky to make a carbon atom into an ion, and it takes many involved reaction steps to force one into an ionic form—a form in which it remains for only an extremely short time. Carbon is partial to covalent bonds.

Three. For the same reasons that it is difficult to make an ion out of a carbon atom, it is rare to find a polar carbon. The proton count in a carbon atom's nucleus—at four—is not great enough to effect an uneven distribution of electron density when it bonds with other atoms. And it is not small enough to allow another atom to draw electrons away to effect uneven distribution of charge. In other words, carbon is dependably neutral in the ways that it shares electrons. Even when bonded with other polarized atoms, it holds them in their hypercharged place without taking on extra polarity itself.

Four. One of the ramifications of carbon's middling mildness is its stability in the covalent bonds it shares. Carbon is steady and consistent in the way it contributes its electrons to its covalent bonds. In fact, this dependability is so great that once a carbon-carbon covalent bond is formed, it takes a huge amount of energy to break it.

Five. In addition to being mildly and dependably reactive, carbon is also promiscuously available for bonding. Since it has four electrons in its second shell, and thus four half-empty orbitals, it can form bonds with one, two, three—up to four other atoms. Carbon, it seems, can be right in the middle of things.

Six. Given that each carbon atom can bond with one, two, three, or four other atoms, molecules created with carbon can take many different shapes. The multiplicity of molecular shapes made possible by carbon— "linear chains, branched chains, and cyclic structures" (Nelson and Cox 2008: 13)—permits specificity in the reactions that form and take place between different molecules that have carbon atoms as their backbone. And specificity is important if a very particular reaction or transformation needs to take place.

Seven (blending one through six). When carbon atoms bond with one another, they form a stable, durable, many-faceted scaffolding for other more reactive atoms and molecular groups. Depending on the kinds of covalent bonds between the carbons in such scaffolding, they can bend, fold, twist, or remain straight to create angular, circular, spiral, or linear molecules. Depending on how carbon is bonded with other atoms—and what those atoms are—it can facilitate, prevent, mediate, and provide a platform for many different reactions. In organic or biological molecules, carbon atoms anchor many reactive or functional groups of atoms so that at the right time, in the right place, with the right molecule before it, a particular reaction can take place.

When we conceive of energy taking form as matter through its constrained self-relation, we see that it is precisely the limitations on how energy can relate to itself that are generative of the different elements, the variety of reactions they undergo, and the proliferation of molecular formations and transformations. And this is a point I want to emphasize. It may very well be the case that at the quantum level, as Barad explains, "ontological indeterminacy, a radical openness, an infinity of possibilities, is at the core" of the processes through which energy takes form as matter (Barad and Kleinman 2012: 214). But at the atomic, cellular, and organismic level, the constraints through which energy relates to itself make it congeal in fairly stable form, not to an extent that it never changes—because it does, and often—but rather in such a manner that we cannot really say that there is a "radical openness" or "an infinity of possibilities." The constraints are the condition of possibility for the formation of atoms, molecules, and cells. Lena Gunnarsson (2013) remarks that "there is nothing about dynamism as such that is at odds with structuredness" (8). And indeed, the amazing and turgid dynamism we observe in the processes through which living creatures persist is not particularly or necessarily unpredictable, unruly, and indeterminate (8–9).

The constraints on energy that make it take form as matter are such that molecules, cells, organs, and creatures cohere in provisionally stable or coherent ways, in ways that make them a somewhat discrete one thing rather than another. Indeed, as I will show in the next chapter, molecules and cells cohere in such a way that the "gappiness" between atoms and molecules that is produced through the tense interrelation of energy creates a porosity or permeability to cells and to flesh. Such coherence, then, is the condition of possibility for a form of discreteness or distinction that comes not from the closure of a boundary but, on the contrary, from a continuous traffic, a constant influx and efflux of molecules in and out, back and forth. The reason I highlight this point is to suggest that such coherence enables us to talk about molecules and cells *interacting* without reinstituting the binary division between "cell" and "environment" or "body" and "world" that this book intends to disrupt. The form of coherence and distinction we see constituted in cells does not trace the logic of identity-as-self-sameness, nor does it, in failing that logic, collapse the possibility of distinction altogether (creating what Timothy Morton describes, in one of his moments of evident frustration, as "the formless goo of Spinoza" [2011: 179]). To think about matter in terms of the constraints on the energy that composes it is to be able to imagine how a form can be discrete even as it absorbs what surrounds it and diffuses elements of itself into its environs.

TWO

Membranes

*In which we learn that the distinction of an
organism vis-à-vis its habitat is a function of
activity rather than substance*

In the previous chapter, I examined how constraints on the way that energy relates to itself constitute what we experience as matter. In this chapter, I will explore how the tense disposition that makes energy take form as matter is the condition of possibility for the movement and transformation of matter. More specifically, I will explore how the energetic vibrancy of matter is bound up with the vitality of living organisms. Critical to this exploration are the membranes that surround, enclose, and define a cell. Brooker et al. (2008) contend that cells are the basic or fundamental unit of life (61). The reason cell membranes are critical is that they are constituted through a very peculiar form of tension between molecules and they are the means by and through which innumerable molecules travel into and out of cells.

When I started studying cells and cellular activity, it became obvious very quickly that the membrane surrounding a cell is not an impermeable wall demarcating and defending an inside from an outside. And this surprised me. My conceptual habit was to think in conjunction with theories of psychoanalysis and deconstruction according to which a violent repudiation, abjection, or exclusion must be effected, monitored, and continually bolstered in order to articulate and maintain a coherent and bounded distinction between inside and outside, me and not-me, self and other. What was so intriguing about cells was that the membranes that enclose and define them are permeable. They are permeable both in the way they are composed and in the peppering of their surface with innumerable channels, gates, and pores that facilitate and force a continual traffic of molecules into and out of cells. Indeed, that traffic is so continuous and so necessary to the activity and survival of a cell that the function of the membrane as a defining boundary became conceptually fuzzy: what kind of boundary is it if it is constituted so as to enable the continuous influx of molecules from the putative "outside" and the continuous efflux of molecules from the putative "inside"? Clearly, as Hannah Landecker (2011) also notes, the distinction between the inside and the outside matters; and the cell membrane has a role to play in fostering that distinction. But the distinction between inside and outside does not depend on keeping the inside and outside separate. How are we to conceptualize the distinctness of a cell when the membrane that encloses it fosters the intermingling and interchange of the substances inside it with the substances on the outside?

Evelyn Fox Keller's work examining the history of genetic sciences was helpful in pushing me to see that my sense of puzzlement about membranes rested on a contestable presumption that the distinction between the inside and the outside of the body is a substantive one. Keller (2010) points out that as Darwin's evolutionary theory was taken up by breeders and agricultural scientists eager to investigate and exploit species changes as they manifest at the level of the individual, they displaced the reigning sense that the distinction between the body and the environment is a function of time, development, and growth, that is, the generative interaction between a living body and the vagaries of the social and material habitat. As these early scientists of heredity endeavored to identify precisely which proportion of physical mutations might be attributable to changes in genes and which to changes in the quality of the environment, the idea that living bodies develop, grow, and change through their engagement with their habitats was displaced by notions of a hereditary biological substrate that is obdurately distinct from the context of its enduring persistence. Keller observes that this effort quite literally to isolate genes compelled scientists to turn away from a temporal framework for figuring embodiment, a framework oriented toward thinking about constitution over a lifetime, and to adopt a framework "that cuts along a different axis, a division between internal and external, and concomitantly between the kinds of substances belonging to these different spaces" (22).

Keller's work showing how the internal realm of living bodies came to be figured as composed of a distinct substance gives us critical leverage. It enables us to appreciate that while a cell membrane does institute a distinction between the inside and the outside of a cell—between body and environment—the porosity and permeability of that membrane mean that the distinction is not a substantive one. As I will show in the pages ahead, the spatial distinction that is marked and constituted by a cell membrane is a chemical or energetic one. A permeable cell membrane produces a continuously variable chemical or energetic imbalance between the inside and outside of the cell, a disequilibrium that in turn creates the conditions for the movement, flow, or dispersion of molecules and their transformation from one kind into another. Such mobility and transformation constitute the processes through which living cells persist in their living. Which is to say that cell membrane permeability is what enables cells and organisms to endure as distinct loci of distinct activities.

In order to gain an appreciation for how the permeability of cell membranes entails a refiguration of the topoi of the inside and the outside, I will begin by examining how the strained and tense relation between different forms of energy facilitates its transformation and movement through chemical reactions and the process of diffusion. Next, I will explore how the turgid energetic détente in molecules between attraction and repulsion, between drawing together and blowing apart, constitutes cell membranes. In the third section, I will look at the kind of distinction that is effected by cell membranes. And in the last two sections, I will explore how the constitution of membranes as porous enables cells to harness chemical reactions and diffusion not only to persist in their activities but also to respond to and to perceive their environments.

Changing Configurations of Energy and Matter

In this section I am going to explore the mechanisms through which energy in the form of matter shifts its configuration, for such transitions are involved in the transformation of matter from one kind into another. The two forms of shift or transition I will examine are chemical reactions and diffusion. As we work through these two forms of transition, it will be helpful to remember that when energy takes form as matter, the electrons that are partially responsible for that material form do not merely spin happily in their orbitals. Rather, they persist in tension with one another. It is this tension that gives dynamism to molecules.

If you recall, electrons, conceived as waves of negative energy, are arranged around a proton-filled nucleus in fields called orbitals. When the electrons spin through these orbitals, they do so very quickly. As electrons spin in a molecule that is composed of several or many atoms, they periodically approach other electrons in nearby orbitals, and then also periodically remove themselves from the vicinity of those orbitals. Because of the mutually repulsive force the electrons' negative charges exercise on one another, the electrons' periodic approach and removal forces the atoms in the molecules to bend slightly away from one another (as the electrons approach) and also spring back (as the electrons retreat). Because the electrons spin extremely quickly, the atoms in molecules also bend and spring back quickly. As the various limbs or branches of molecules bend and spring back, over and over again, they seem thereby to vibrate.[1] This vibration is critically important, for it affects the way molecules interact with one another when they get close to one another.

When vibrating molecules come into proximity with one another, the vibrancy of each constrains the vibrancy of the other: their individual patterns of vibration limit each other's bend and spring movements. This limit makes their behavior vis-à-vis one another more ordered, in the sense that they do not vibrate and jiggle around as much. But in restricting that movement, this limit also generates a frenzied and unsustainable tension in and between the molecules, for the electrons involved end up being too close together as they spin in their orbitals. Since the frenzied tension of proximity is unsustainable, the electrons either shift their configuration to create a bond between the molecules or they repulse one another so as to propel the molecules away from the constraints of that proximity.

Let's take the bonding first. When the tension produced through constrained vibration reaches a screaming height—or a threshold—the electrons in the molecules involved are exchanged or shared to create either an ionic or a covalent bond. In so doing, they form a new molecule, that is, a different form of matter. These kinds of exchanges or sharing of electrons are chemical reactions. In other words, chemical reactions are changes or transitions in energy, interactions between atoms and molecules in which the energetic tension produced by their proximity is released or diminished by a joining, by a bond. With such a bond, the energy level of the molecules involved diminishes; the reaction and its rearrangement of electrons and of the regions that vibrate enable the now-conjoined molecules to be differently constrained, less constrained, and thereby more stable.

Diffusion is another possibility when the proximity of atoms or molecules constrains their vibration. Here, the intensity of vibration that increases as the molecules come close together reaches a pitch such that it pushes them apart. As the molecules repulse each other and move further away, the energetic tension generated by their proximity dissipates. As the molecules disperse through space in their energy-dissipating way, the patterns of their interactions with one another become more erratic and disorderly. Diffusion, then, is the movement of similar molecules away from one another such that their relative concentration in a particular spatial location gradually decreases. For example, when someone wearing cologne walks into a room—let's say a large room—within minutes of his or her arrival most of the room's other occupants will have been treated to a whiff of the scent he or she is wearing. What is always striking about such a situation is how very quickly the odor travels and how thoroughly it occupies the full extent of a room's real estate. The spreading, thinning, dissipation of the scent is the work of diffusion.

What is important to recognize in this spreading, thinning, and dissipation is that diffusion always occurs down a concentration gradient. What this means is that the movement of dissipation is always from a region of higher concentration to a region of lower concentration (hence the scent spreads). The reason that molecules diffuse down a concentration gradient centers, again, on the fact that their vibration and collision with one another makes them energetically "tense," and their mutual repulsion and movement away from one another enables them to move more erratically and freely such that they become energetically "relaxed." So, molecules tend to shift from higher to lower energy states; diffusing molecules shoot, bump, and drift about until they reach the most disordered and least energized state allowed by the context in which they exist. Note that it is not the case that once in the most disordered state molecules stay in any one particular spot: at the lowest possible concentration, they move around randomly and speedily. But in that disordered state, there ends up being an even average distribution of molecules as they whiz about.[2]

Chemical reaction and the process of diffusion are two means by which the molecules that compose matter transform. Indeed, reaction and diffusion are the two primary modes through which the matter that composes organisms perpetuates living activity. And what is crucial to the perpetuation of these reactions and diffusions is the cell membrane. On the one hand the cell membrane creates an enclosure that compels molecules to be closer in proximity than they might otherwise be; the membrane prevents wanton diffusion and even corrals molecules so that they approach one another and more easily reach the energetic tension thresholds necessary for a reaction to take place. On the other hand the cell membrane is constitutively porous—as it must be if the forces of diffusion are to replenish the molecules that are used up in reactions and remove the products of those reactions so that the reactions can continue to occur. And while chemical reactions are what create the molecules that compose a cell membrane, the forces of diffusion are what hold those molecules together in membrane form.

Membrane Formation

One of the defining features of a cell is the membrane that encloses the cell and constitutes a boundary between the inside and the outside. And what is fascinating about cell membranes is that they form when diffusion provokes the self-sorting of different kinds of molecules and their separation from one another. There are two things I want to highlight as I move into the explanation of membrane formation. First, we often conceive of boundaries as constitutive of an inside through the demarcation and exclusion of an outside. The idea here is that since there cannot be a definitive inside without an outside, the boundary performs the necessary constituting exclusion. However, in the case of cell membranes, we see a peculiar twist of the usual conceptual logic: the boundary that demarcates the inside and the outside is constituted through the separation and seclusion of the molecules that compose the membrane. In other words, the boundary, and not the outside, is the excluded term. Since neither the inside nor the outside are excluded, then we can anticipate or expect to see ongoing interaction between them—which we do.

The second thing I want to highlight is that the molecules that together compose a cell membrane do not bind or grip or fuse to one another. They do not join together in a bond. Rather, the molecules jostle side by side in liquid. Indeed, they huddle together to avoid the liquid that then becomes both the content of and the context for their formation. As is suggested in this notion that the molecules composing a membrane huddle, this grouping or conglomeration does not produce a sealed, impenetrable border. Rather, it generates a loose boundary through which all kinds of other molecules might seep and flow. In other words, since the molecules composing a cell membrane huddle rather than bind to create hermetic seal, the forces of diffusion can push molecules into and out of cells.

To understand how this huddle happens, we need more details about how diffusion occurs in different kinds of molecules.

Some carbon molecules form in a long linear chain, with each of the carbons in the chain covalently bonded to one another while also bonded with (holding hands with?) two hydrogen atoms. Such chains end up looking a bit like caterpillars or centipedes with lots of legs. This kind of chain is a fatty acid or a lipid molecule.[3] Because fatty acids are long stable carbon chains with lovely stable covalent bonds both between the carbons and between the carbons and their hydrogen appendages, they are rather chemically inactive. There is not a lot of excess or misplaced energy trolling around fatty acid molecules, and they tend not to interact with other molecules. Indeed, their energetic quiescence makes them a nuisance to molecules that are energetic or that are composed in such a way as to have an energetic imbalance. Molecules like water, for instance.

A water molecule is composed of one oxygen atom with two hydrogen atoms attached. We often refer to water as H_2O, which is its chemical formula, but for the uninitiated this notation could imply that the two hydrogens are attached to each other. A better way to write it is as HOH, for this gives a more accurate sense that the oxygen is the central and anchoring atom for the molecule. But, importantly, in contrast to the nice neat line formed by the letters on the page, the atoms that together compose water molecules do not form in a straight line.

As noted in the previous chapter, oxygen has six electrons in its outer electron shell, which means that it needs two more to make a stable eight. Since electrons spin by twos in orbitals, four of those six are already paired in their orbitals. The remaining two unpaired electrons in oxygen's outer electron shell each occupy half an orbital and await a pairing through the formation of a covalent bond. Now, in the case of water, those two unpaired electrons bond with hydrogen atoms. So picture this: the pair of already-paired electrons exerts a repulsion on the other electrons in the oxygen atom, a repulsion that maintains its effect when the unpaired electrons bond with hydrogen atoms to create a water molecule. As a result of this repulsive action, a water molecule is V-shaped, like a boomerang, with the oxygen at the apex in the middle and the two hydrogens on the wing on each side (Nelson and Cox 2008: 43).

In addition to being V-shaped, a water molecule is polar. Because the oxygen atom in the molecule has more protons in its nucleus than do the hydrogen atoms, the oxygen pulls the bulk of the electron energy in the molecule toward it. The success of the oxygen in its greediness for electrons ends up creating an unbalanced distribution of electron charge throughout the water molecule: the oxygen at the apex of the V ends up being a relatively negative side of the molecule, and the hydrogens on the wings tend to be slightly positively charged.

Both the shape and the polarity of water molecules contribute to its behavior when energy is added to or subtracted from it in the form of heat. At very cold temperatures, water molecules vibrate hardly at all. As a consequence, they form a three-dimensional lattice with each other. Here, the more positively charged hydrogens in the water molecules create hydrogen bonds with the negatively charged oxygens in other water molecules. Together, in their slowly vibrating, low-energy lattice structures, they compose what we know as ice (Nelson and Cox 2008: 44). Liquid water is fluid because the extra energy or heat in it makes the molecules vibrate more such that the hydrogen bonds connecting the individual molecules are readily made, broken, and remade. The transience of the hydrogen bonds in liquid water means that the molecules float and slide around each other quite easily. Vapor water or steam is airborne because the addition of further heat creates vibration in the molecules to such an extent that it overwhelms the hydrogen bonds: the water molecules individualize and float away (43–45).

Now, let's mix molecules.

When a substance composed of ions, like salt or sodium chloride, is mixed in water, the polar water molecules surround the sodium and chloride ions, interrupting the electrostatic forces of attraction that hold the salt into a visible solid. The water molecules slowly peel the ions from the solid and carry them away. This is what it means to say that salt dissolves into a saltwater solution. When all the water molecules in a given volume of water are already occupied with either sodium or chloride ions and there are no further water molecules available to surround and dismantle the salt, then the solution is said to be saturated—no more salt can be dissolved in it.

In contrast with the example of salt, if a fatty acid like oil is added to water, it cannot dissolve into solution. As noted in the previous chapter, covalent bonds between carbons are very strong. And in fatty acids, so even are the bonds that hold the atoms in their molecular formation that there is no relative difference in the distribution of charge around the molecule. What this means is that, unlike water or salt, a fatty acid molecule is nonpolar. Because of their energetic stability, fatty acid molecules do not even transiently bond with water molecules that circulate around them. In fact, in not being able even transiently to hydrogen bond with the polar water molecules, the nonpolar fatty acid molecules interfere with the interactions between the water molecules themselves. The fatty acid molecules are so energetically staid that they get in the way; the water molecules cannot tumble easily around them. Unable to vibrate and shift around freely, the water molecules are so constrained in their fluid formation and reformation of hydrogen bonds that they end up forming something of a rigid structure around the offending insoluble fatty acid molecules. The result is that when oil is added to water, the water molecules end up being more ordered, more constrained in their vibration and movement, and energetically extremely tense.

This is a complicated dynamic, so try to think of it in terms of how individuals holding hands might move through a crowd. Imagine individuals who may move through a crowd only by holding hands in pairs. Individually, their travel is faster if they can let go of one person and grab onto another. This scenario leads to a freewheeling, fairly chaotic set of interactions between lots of individuals. However, if some of the individuals in the crowd refuse to hold hands, that refusal means that those individuals who want to let go in order to grasp a new hand cannot let go: they are stuck in a pair until they can find other willing hand-holders. Because they are stuck in a pair, moving through the crowd becomes more difficult and consequently slower—the agitation and frustration of the forced-to-stick-together hand-holders grows. This is akin to what goes on when nonpolar fatty acids are added to water.

Now, because such higher energy states are unstable and because higher energy states tend to devolve into lower energy states, something very cool happens: the nonpolar fatty acid molecules cluster together—closely—in such a manner as to lower the surface area that is exposed to the frustrated water molecules. (Imagine the hand-holding refusniks gathering together into a group in order to limit their encounters with the grumpy hand-holders.) This clustering together of nonpolar molecules in water is called a hydrophobic interaction. When the hydrophobic molecules cluster in this way, water molecules are released from their frustrating constraint and so can freely disperse.

Hydrophobic interactions are the reason that oil and vinegar separate in your bottle of salad dressing: the oil (nonpolar) and vinegar (polar) separate into two distinct layers so that only the layer of molecules at the line of separation are touching and the rest of the molecules can move freely. And when you shake the dressing to blend it prior to pouring, you are creating a highly frustrated and energetic—if delicious—mixture for your salad.

Fixate for a minute on the clean definitiveness of the line in your salad dressing bottle that marks the separation of the oil and vinegar; observe in your mind the brisk return of the little globules of oil and vinegar to their proper places after you have shaken the bottle. The reason this is important to focus on is that the process of cluster and separation in hydrophobic interaction is the very same process through which membranes are formed—membranes such as those we find in cells.

But to appreciate membranes in cells, we need first to understand another kind of molecule. Although I have focused so far on polar and nonpolar molecules, there are some molecules that are neither wholly polar nor wholly nonpolar. Instead, they have one end or region that is polar and another that is nonpolar. These are known as amphipathic molecules (think: amphibian, living both in water and on land, but the "both" in amphipathic refers to the quality of the "pathos"—they are structurally ambivalent in their feelings about water). Amphipathic molecules interact with water in a very peculiar manner, for one end, hydrophilic, can mingle easily with water molecules, while the other one, hydrophobic, cannot. When added to water, amphipathic molecules tend to do one of two things.

One option entails that they form small spheres (visualize a very full pin cushion) at the center of which are hidden the nonpolar hydrophobic "tails" and on the surface of which are gathered the polar hydrophilic "heads." These spheres are called micelles. The role that diffusion plays in the formation of micelles is as follows: the arrangement of the amphipathic molecules in micelles sequesters the nonpolar ends of the molecules from the water while also allowing the polar ends to interact with the water. Both of these developments contribute to the least energetically tense form of interaction for the substances: all the players are happy. Incidentally, micelles are what make detergents work: they embrace blobs of dirt, oil, or grease within their hydrophobic center and spirit them away through water—leaving your dishes sparkling clean.

In the second option, rather than enclosing themselves into a solid sphere with tails in the center, the amphipathic molecules gather together in two layers, with the polar heads on the outside and the nonpolar tails hidden sandwiched in between. As this sandwich spreads or grows, it forms a membrane whose lowest energy formation—the one in which there is the least contact between nonpolar tails and water—is one in which the edges of the sandwich curve around and connect to each other. The result is something like a tennis ball made out of a thin sandwich of fabrics: hollow on the inside—an inside that is enclosed and closed off from the outside. Voila! One of the basic constituents of a cell. In a cell, such a membrane is called a lipid bilayer membrane—let's repeat the term: a lipid (an amphipathic molecule with a fatty nonpolar tail and a polar head) bilayer (two layer) membrane.

And what is so amazing and indeed incredible about the lipid bilayer membrane from which all our cells are composed is that the molecules are held together via hydrophobic interaction. Those molecules hold themselves together not because they fit, grip, or grab in some way, not because they are attracted to or form bonds with one another, but rather because their being side-by-side liberates water molecules to become freely dispersive. The amphipathic molecules adhere to one another as a result of the concerted absence of water along the surfaces that touch, by the physical fact that water has a tendency to avoid such energetically "tense" situations and is energetically more "restful" when it can avoid them.[4] A hydrophobic interaction does not seem like much when talking about the construction of cells. It seems as if there should be some kind of glue or grip or fastener binding and holding everything together. But hydrophobic interactions are clearly sufficient because, well, cells happen and life goes on.

A Strange and Unfamiliar Distinction

The fact that amphipathic lipid molecules are held together in membrane form merely by hydrophobic interactions is—for a cell—a perfect accident of the energetic constraints that form and reconfigure matter. The activities that occur within and between cells are all essentially chemical reactions, interactions between molecules that form bonds, disintegrate, transform, and then interact with other molecules in different ways. Chemical reactions deplete the concentrations of some molecules as they transform them; correspondingly, they increase the concentrations of those molecules that the reactions produce. As these reactions take place over time, then, they both require and produce shifting concentrations of different kinds of molecules. The repetition of those reactions relies on and provokes processes of diffusion across the porous cell membrane. Generally speaking, if a reaction inside a cell uses up a specific kind of molecule, the concentration of that molecule will be lower inside the cell than on the outside—which will prompt a diffusing redistribution or shift of those molecules into the cell until the concentrations are more or less balanced again. Similarly, if a reaction creates a higher concentration of a particular kind of molecule inside the cell, the forces of diffusion will push and rearrange the molecules across the cell membrane until the concentrations are roughly equivalent. The cell's biochemical activity relies on and provokes the processes of diffusion—and it is the cell membrane that provides the conditions for those processes to be quite precise.

To see how and why diffusion takes place across or through a cell membrane, let's look at cell membrane composition in more detail.

Seen from above, a cell membrane looks something like what the surface of your bathwater would look like if you were to throw in a hundred or so ping-pong balls to float there: a bustling, shifting mosaic of molecules—in this case, lipid molecules. The reason the ping-pong ball analogy is useful is that it enables us to visualize the fluidity of the cell membrane. Just as the imagined ping-pong balls float and bustle around each other—and permit the rubber ducky to meander slowly about—so the lipid molecules shift and bustle around one another, with protein and cholesterol molecules either drifting around among them or creating anchor points that circumscribe that flow. In fact, Brooker et al. (2008) note that at human body temperature, lipid molecules in cell membranes switch places with each other around 10 million times per second—which is a very busy bustle indeed (89).

Significantly, then, the cell membrane is not a seamless and solid wall of molecules. Instead, as Nelson and Cox (2008) explain, it is a "fluid mosaic" (373). The fluidity of the cell membrane suggests that there could be a fair influx and efflux of molecules across the membrane. Indeed, the cell relies on and exploits this porosity. There is a constant and continuous diffusion and traffic of substances into and out of cells. However, not just anything can pass through the cell membrane. As Brooker et al. (2008) note, a cell membrane is constitutively leaky but not indiscriminately so (89). Rather, cell membranes are selectively permeable, which means that they are constituted to enable the flow of select molecules in and out of the cell as required by different reactions.

Some of this selective diffusion is a function of the energetic constraints that create cell membranes. Which is to say that for some molecules, the composition of cell membranes is such that they cannot but enter and leave as the forces of diffusion wax and wane. Molecules that are especially small, like nitrogen, oxygen, or carbon dioxide, can squeeze between lipid molecules—think here of a marble plopping through the floating mass of ping-pong balls. Small fatty acid or lipid molecules, like steroids or some hormones, can also seep through: they are nonpolar, just like the lipids that compose the membrane, and so can join the hydrophobic bustle that constitutes the membranes and then slide through to the other side. Such small and nonpolar molecules can move back and forth across the cell membrane via simple diffusion.

However, polar molecules like water, charged ions, and larger molecules like carbohydrates or proteins are unable to squeeze a passage between the lipid molecules that compose a cell membrane. So, when a chemical reaction requires a concentration of such molecules—as they often do—the cell will provoke a series of reactions that produce proteins that are tube-like, channel-like, or gate-like. These tubes, channels, and gates become embedded in the cell membrane and facilitate the movement of those needed molecules into or out of the cell. In other words, the diffusion of molecules back and forth across the cell membrane is not merely an accident of the ways that the lipid molecules hold together. The cell accommodates its need for those molecules by building and putting into place in the cell membrane special entryways and exits.

Many of these pores, channels, and gates enable specific target molecules to traffic in and out of the cell as provoked by the forces of diffusion. For example, an aquaporin is a pore that lodges in the cell membrane and enables water molecules to pass through the nonpolar lipid core of the membrane according to the pressures of diffusion. A sodium ion channel is a channel that specifically enables—yes—sodium ions to pass. And so on.

But whereas some pores or channels are simply open and thereby enable specific molecules to trickle or wriggle through, others are more obviously selective in the sense that there is a gate that physically opens and closes. Typically, in such gates, a target molecule slips into a pocket on the gate that fits its specific shape and electrical charge. On slipping into place, the molecule changes the resonance of the molecules that compose the gate, which triggers a bend or a change in shape of the gate such that the gate shifts the open portion of the pocket to the other side of the membrane. This shift, or change in shape, transforms both the charge and the shape of the pocket such that the molecule no longer fits: the molecule drops out, which in turn reconfigures the gate so that it resumes its original shape (Brooker et al. 2008: 98). The best analogy here is the way that a gumball machine takes in or accepts coins: there is a flat round opening for the quarter coin, you put it in, and turn the handle. The presence of the coin pushes a mechanism that allows the handle to rotate fully so that, on the inside of the gumball machine, the quarter drops out of the quarter slot and into the coin-holding box.

The gumball machine analogy works especially well to explain a piggyback process called co-transport, in which a gate that opens for one kind of molecule also facilitates the travel of another kind of molecule. It is not hard to imagine that the turning of the gumball machine crank that deposits the coin also turns a crank that results in the ejection of the desired gumball. The case of co-transport in a cell is similar. For instance, when the gate-pocket from the outside appears on the inside (thus precipitating the deposit of the entering molecule), a different pocket is exposed in the gate, a pocket that a specific exiting molecule can occupy. When the entering molecule drops off the transporting gate, triggering a change in the shape of the gate, the second opening cradling the exiting molecule shifts to the outside of the cell. In this configuration of the gate, it is impossible for the exiting molecule to remain in the pocket—which is to say that the change in charge and shape causes the molecule to fall out. Co-transport gates are used by the cell to enable molecules to stow away on a diffusing molecule's trip, traveling either with or against the direction of that diffusing molecule (Brooker et al. 2008: 99). As I will show in chapter 4, co-transport can be used to compel chemical molecules to migrate into or out of the cell against their concentration gradient, that is, moving from a lower to a higher concentration. In other words, co-transport gates can make the passive diffusion of molecules contribute to the ratcheting up of the concentration of other molecules so that a reaction can take place.

So, in addition to being composed in such a way as to enable the diffusion of molecules in and out of the cell, cell membranes are teeming with gated channels and pores that ensure that the molecular traffic in and out of the cell makes possible the changes in concentration that are the condition for various biochemical reactions that the cell must undertake. And those reactions are various. Nelson and Cox (2008) observe that each cell generally has "more than 50 kinds of plasma membrane proteins" (386) that work to facilitate the transit of a wide array of chemicals across the cell membrane. And lest one think that such channels must make molecular traffic a terribly slow and congested affair, Brooker et al. (2008) note that "when a channel is open [and working], the movement of molecules across the membrane can be extremely rapid, up to 100 million molecules per second" (97).

If the possibility of chemical influx and efflux is exactly what is ensured by the way the cell membrane is composed, then the way we think about that infusion and effluence will not be in the terms of a breached or failed boundary. For a cell, influx is not an invasion; efflux is not a loss. Instead, both movements are absolutely critical to the continued biochemical activities through which cells persist over time. The boundary that is the cell membrane exists not to prevent entry and exit but rather to allow and ensure them. Accordingly, we shall have to suspend our sense of a cell membrane as a marker, maker, and enforcer of substantive difference. Instead, we shall have to think of a membrane in terms of the chemical transitions and transactions it makes possible.

So although our conceptual habits often tell us the contrary, the distinction between the inside and outside of a cell does not map onto a distinction between the "matter" of biological substance and the "matter" of the environment. We can think about the permeability of cell membranes in terms of Stacy Alaimo's (2010) concept of transcorporeality, which she says "indicates movement across different sites" (2). If we think of cell membranes in terms of the movements and shifts they make possible, then the substance of the inside and the substance of the outside become essentially indistinguishable. As Alaimo puts the point, "'the environment' is not located somewhere out there, but is always the very substance of ourselves" (4).

A cell membrane provides the conditions for chemical distinction; it marks and creates a three-dimensional zone within which a distinctive set of activities proceeds. As a porous division, it makes possible the influx and efflux of chemicals, the rise and the fall of concentrations of molecules, that are so necessary to a cell's biochemical activities. Since precisely those biochemical activities are made possible by the traffic of molecules across the membrane, we could say that the biochemical processes that occur within the permeable enclosure of a cell are what distinguish a cell from its environment. What makes a cell distinct are not the substances that lie on either side of its enclosing membrane but the myriad biochemical activities that occur within it and on and through that membrane.

What is particularly helpful in this conceptualization of what makes a cell distinct is that it allows us to take account of the ways that a cell is fundamentally constituted with and by means of its environs while also acknowledging that a cell is differentiated from those environs in such a manner as to be able to respond to, perceive, and engage with them.[5]

Response and Perception

To conceive of the cell's distinction in the way I have outlined here provides the occasion for us to appreciate how the porosity of cell membranes creates the conditions of possibility for a cell's responsiveness to its environment. As I have shown, because energy is constrained in the ways it can relate to itself, a cell membrane is composed in a manner that permits and facilitates the continuous, turgid, and variable flow of molecules into and out of a cell. But as Nelson and Cox (2008) point out, a cell does not passively vehiculate—take in, pass out—whatever the currents and flow of matter might be in its immediate surroundings (371). The chemical changes and energetic transformations effected by the molecular traffic across the membrane provoke chemical reactions inside the cell that either take advantage of or inhibit the effects of the influx and efflux of chemicals. Some of those chemical reactions provoke changes in the selectivity of the membrane's permeability (for instance, increasing or decreasing the number and kinds of channels, pores, or gates), and some amend incoming molecules so as to enhance the speed at which they flow. In other words, as the chemical composition of a cell changes through the chemical reactions that occur in it as well as through the processes of diffusion, the cell responds to the quantity and kind of molecules that stream in. As a consequence of the varied but constant traffic across the cellular membrane, a cell is responsive to its own activity as well as to its environment.

I have already shown how cells create pores and channels to enable and facilitate the movement of different kinds of molecules across the cell membrane. This creation is an ongoing process that, by means of a very complex network of chemical reactions, is responsive to the conditions required for the cell's biochemical reactions. So, for example, if the number of a particular kind of pore or channel is insufficient to provide for the concentration of a specific kind of molecule necessary to trigger a chemical reaction, then the cell will produce more of those pores or channels. The sequence runs roughly thus: either the low concentration of the specific required molecule or the concomitant low concentration of the products of the chemical reaction creates a cascading series of other reactions, each individual one of which is the provoking condition of the next. This cascade of other reactions eventually affects molecules inside the cell's nucleus, provoking them to make more of that particular kind of pore and then to shuttle them to the membrane to embed them so they can do their work of making possible chemical traffic.

In addition to creating more of or dissolving a portion of specific pores, channels, or gates embedded in its permeable membrane, a cell adjusts the flow of chemical molecules entering the cell by making a small amendment to the molecules before they are used in reactions. A terrific example is the tagging of glucose (sugar) molecules as they enter cells via glucose transporter gates. Glucose molecules are used quickly in cellular reactions, and the inflow needs to be fairly constant. However, simply to have glucose molecules flow into the cell just so would create the conditions for a flow slowdown: the just-in glucose molecules might prevent the not-yet-in ones from diffusing in because of a high local concentration near the glucose transporter gates. However, when glucose molecules first enter a cell, they are tagged in a little chemical reaction—tagged with something called a phosphate molecule. These tags lower the concentration of "pure" glucose molecules and thus enable the diffusion in to continue. I know. Tricky. Ingenious.

Since the cell membrane is constituted in such a way as to enable the cell to respond with fairly exquisite precision to shifting concentrations of molecules in its immediate environs, we could say that the living cell is a perceiving cell.[6] This is a particularly compelling possibility when we consider that the molecules that flow out of cells affect the concentrations of chemical molecules that neighboring cells absorb and to which they in turn adjust. But to be more pointed about this, while all cells selectively incorporate and evacuate a variety of biochemical molecules, there are some cells that harness permeability and diffusion to affect other cells in ways that enable a multicellular organism to absorb and respond to the world *as an organism*. These kinds of cells are called nerve cells. Now, while Stephen Rose (2005) provides a wonderful account of how nerve cells might have developed over time, I am going to use the permeability of the cell membrane along with insights about constraints on energy's self-relation to explain how they work.

Nerve cells enable a multicellular organism to absorb the world and respond as an organism to it. They do this by absorbing or interacting with the environment in ways that change the local chemistry in the interacting cells. This chemical change diffuses or translates through a chain of nerve cells in such a way that other cells in the organisms adjust their activity so that the organism can both respond to and survive the encounter.

Nerve cells are distinguished from others in that they are typically very long: they have a somewhat rounded cell body with short appendages on one end, called dendrites, and a long, stretched-out tail on the other, called an axon. Because nerve cells are so long, the influx of molecules at one point on the cell membrane may not diffuse immediately throughout the entire cell; sometimes the influx and its effects remain quite local. Indeed, the very localness of diffusion across a nerve cell membrane enables a nerve cell to transmit an electrochemical signal along its length to another nerve cell and eventually on to the central nervous system. Let me explain what it means to say this.

The membranes of nerve cells are peppered with channels that selectively enable sodium and potassium ions to travel in and out of the cells. On the one hand the traffic can be tracked in terms of the chemical gradient, that is, the differential between the concentrations of the sodium or potassium ions on one or the other side of the membrane. But since the sodium and potassium ions are charged—positively charged—the traffic also has an effect on and shifts in accordance with an electrical gradient. So, as chemical diffusion pulls the ions in and out of the nerve cell, the electrical gradient across the cell membrane changes. In turn, the electrical gradient facilitates a diffusion of ions that changes the chemical gradient. As the influx and efflux of these ions transforms the cell membrane locally around the channels, those local shifts produce local diffusion that triggers adjacent pores to undertake similar chemical and electrical diffusions across the cell membrane.

The way that the local diffusions shift, spread, and travel along the length of a nerve cell demonstrates how the permeability of the cell membrane is the condition for the cell to respond to and perceive its environs. The illuminating details of that process are as follows.

There is a special sodium-potassium gate in the nerve cell membrane that opens and closes fairly regularly.[7] As it opens and shuts, it shuttles three sodium ions out of the cell and at the same time carries two potassium ions in—a little like the gumball machine alluded to earlier. The activity of this pump produces a chemical gradient across the membrane: there are more sodium ions on the outside than on the inside and more potassium ions on the inside than on the outside. Indeed, it is because of this disparity in concentration that we say that the gate pumps the ions *up* their chemical gradient, moving against the forces of diffusion.

Importantly, the activity of the sodium-potassium gate also produces an electrical gradient across the cell membrane. As can be easily calculated, given the positive charge of each of the ions that the pump moves, "each cycle of the pump," with three positive ions out and two positive ions in, "results in the net movement of one ionic charge across the membrane" (Nichols et al. 2001: 61–63). So, as this gate pumps, its activity creates an electrical gradient across the cell membrane, with the outside becoming relatively more positive than the inside, and the inside becoming relatively more negative than the outside.

Given what we know about the forces of diffusion, we can imagine that if a hole were to open in the cell membrane, sodium ions would diffuse in and potassium ions would diffuse out. Indeed, this is exactly what happens when a nerve cell is activated and distinct sodium ion channels and potassium ion channels open and close to redistribute the chemical and electrical concentrations.

Imagine that sodium ion channels open quickly, close quickly, and then stay closed for a moment before opening again. The rhythm of this opening and closing sequence is: quick, quick, pause. And imagine that potassium ion channels open slowly but stay open for a while before closing again. The rhythm of this opening and closing sequence is: slow, slow. The sodium and potassium channels work in rhythmic sequence together. And like any good foxtrot, the rhythms proceed as follows: quick, quick, pause, slow, slow, quick, quick, pause. So, the sodium ion channel opens quickly, which allows a flood of sodium ions into the cell, and then the channel closes. Then the potassium channel opens allowing potassium ions to pour out, and then the potassium channel closes. In both cases, the ions diffuse down their concentration gradient.

What is interesting here is how the pouring in and then out of the ions changes the local charge across the nerve cell membrane. As the sodium ions pour in, their positive charge diminishes the negative charge local to the sodium ion channel on the inside of the cell membrane. This diminution is called depolarization because the membrane no longer has a polar charge: it becomes electrically neutral. When depolarization occurs, the electrical neutrality slams shut the sodium ion channel and triggers the opening of the potassium channels. The potassium ions pour out, as a consequence of which the local charge around the channel becomes negative again. And because the potassium channel remains open for several long splits of a second, the local negative charge increases and spikes before coming back to its resting or standing level.[8]

Significantly, when the sodium ions diffuse into the cell through their activated sodium ion channel, they then diffuse along the inside of the membrane and depolarize other sodium channels local to the one through which they poured. Nichols et al. (2001) explain that this additional depolarization of additional sodium channels makes them open, with the effect that more sodium ions pour in and depolarize further sodium channels (93). Crucially, because the sodium ion channels stay closed for a bit immediately after having been opened—remember that they pause—the depolarization of the membrane that opens sodium ion channels leaves those that have already, just recently, been opened and closed unaffected (pause) while opening those adjacent to the original spot. If this depolarization of as yet unaffected adjacent spots is repeated over and over again, we can see that the location in the membrane that depolarizes shifts (Nelson and Cox 2008: 410). Indeed, if one imagines starting a depolarizing shift on a nerve membrane on the left side of the page, the pause that is so critical to the sodium channel's rhythm would force the locus of depolarization and repolarization to move moment by moment toward the right side of the page.

This is just like a fan-wave in a stadium, where people stand up and cheer, and then sit down and wait a moment before they heave themselves to their feet again: the pause enables the wave to ripple around the stands. In the nerve cell, the pause in the rhythm of the sodium ion channel's activity enables a ripple of depolarization and repolarization to travel down the length of a nerve cell. Notice that, just as in a stadium wave, it is not the people but rather the ripple of their activity that travels, so what travels along the length of the nerve are not the ions themselves but the patch of depolarization and repolarization on the membrane.

So, when different forms of light enter the eyes, when sound waves vibrate eardrums, when chemical molecules hit taste buds in the mouth or odor sensors in the nose, when flesh is compressed or brushed sharply, heavily, or lightly, when something hot or cold touches or surrounds the skin, or when a body moves and orients itself in relation to itself and to space—the light or vibration or chemical or temperature change or stretch of cell membranes triggers the depolarization of nerve cells such that the change absorbed by the cells creates a cascade of depolarization and repolarization. This chasing cascade of depolarization and repolarization travels the length of the nerve cell. And when it reaches the end—the axon—it causes the nerve cell to spit out biochemical molecules called neurotransmitters, which strike the nearby dendrites of another nerve cell (a connection that is called a synapse) and initiate a wave of depolarization and repolarization through that nerve cell.

The nerves in multicellular organisms go from the extremities of the physical body to the central nervous system, where they extend to form the crazy conglomeration of nerve cells that is the brain. As signals precipitated by sensory stimulation travel to the brain, the system of nerves organizes two kinds of response. One kind of response, in the central nervous system, processes the information and responds by provoking positive or negative feelings along with often deliberate, conscious thought about what those signals say about the world, and by shooting similar nerve signals back down to muscle cells to effect gross physical action. The other kind of response, in the autonomic nervous system, acts to ensure that the cells and the body as a whole can survive and even benefit from the encounter and engagement that provoked and that resulted from the response of the central nervous system. Nerves in the autonomic system do things like adjust the speed of breathing (thereby affecting the amount of oxygen in the blood) and heart rate (thereby affecting the circulation of the blood) and make other adjustments in bodily temperature, digestion, and vision that enable the body successfully to engage the world (Pratt et al. 2008). The nerves in the autonomic system also transmit their signals to the body's endocrine systems.[9]

The endocrine systems are a network of organs and glands that do their work via diffusion and blood circulation rather than via nerves. Endocrine organs produce steroids and hormones of various kinds, chemicals that, as Koeppen and Stanton (2010) explain, "regulate essentially every major aspect of cellular function in every organ system" (662). When they are released into the blood or into tissue in a specific local area, endocrine chemicals either slip through cell membranes or land on the exterior surfaces of cell membranes in ways that trigger reactions on the membranes' other sides. The reactions provoked by endocrine chemicals affect the amount of water and the concentration of sodium, potassium, and other ions in cells; they adjust sugar levels, insulin levels, red blood cell levels, cholesterol levels;[10] and they modulate the activities of the reproductive systems. In doing so, they signal the kinds and quantities of pores, channels, gates, and transporters that the cells need to produce for the membrane so that cells can continue to import and export the chemical molecules necessary for their continued persistence (Becker et al. 2002: 30–31).[11]

Scientists have tended to consider the different networks of autonomic nerves and endocrine glands as distinct axes of activity in the body. However, Viau (2002) and Tanriverdi et al. (2003) contend that cells in many different organs have landing spots or receptors on cellular membranes that can receive steroid and hormone molecules and provoke a cellular response to them. In other words, rather than being limited to a discrete set of reactions and cells on which they can have an effect, steroids and hormones have a regulatory effect on a variety of different biological activities, often working in tandem to fine-tune the body's engagement with and response to its environment (Tanriverdi et al. 2003; McEwen 2001). Tanriverdi et al. (2003) and Webster et al. (2002) explain that steroids and hormones even work in conjunction with the immune system so that they work together to direct the body's response to chemical, bacterial, viral, or social stressors: energy might be directed to an immune response, or an immune response might be repressed so that energy can be directed to address the stimulation of the stress in a different way.

From Porous Membranes to Biocultural Creatures

In this chapter, I have tried to explain how the constraints on the ways that energy relates to itself effect the formation of cell membranes so that they are porous. The porosity of cell membranes is the condition of possibility for the many biochemical reactions that occur within a cell. Via interconnected networks of chemical reactions and processes of diffusion, this porosity enables a cell to respond to its own biochemical activity as well as to changes in its environs. The porosity of cell membranes is also the reason that nerve cells can translate an electrochemical signal along their lengths and to other nerve cells, which is the basis of an organism's ability to perceive, respond to, engage, and survive changes in its body and habitat. So, if we can understand the peculiar work of cell membranes in defining not a substantive distinction between inside and outside the cell but rather a distinction concerning biochemical reactions, we can extend that insight imaginatively to recompose our sense of the outlines of a living organism.

When we think about living bodies, we do not need to imagine that they are wholly and thoroughly a part of their habitats simply because we reject the notion that they could be wholly separate from them. In between the conceptual options of a smear and a hermetically sealed container is the porous body. What makes a living body distinct from its environment are not the substances of which it is composed (which in fact traffic back and forth across the membranes of the body's constituent cells constantly) but rather the activities and the processes that occur within and by means of that body. And given the activities that take place within cells, it is not simply that organisms absorb the chemical and material constituents of their habitats. Rather, since cells build and dissolve the protein gates and channels that then regulate and recalibrate the influx and efflux of molecules (they build and dissolve many other proteins, too, that participate in that regulation and recalibration), bodies quite literally rebuild themselves, constantly, in response to the molecular constituents of their habitats, to the physical features of their habitats, and to various stimulants in their habitats. This is what researchers like Bruce McEwen mean when they claim that the material and social environment "gets under the skin" (2012).

If we can hold on to the notion that biochemical processes and activities are what distinguish the inside from the outside of a cell or an organism, we can apprehend just how profoundly and fundamentally material and social environments get under the skin without at the same time losing the conceptual possibility of talking about organisms or bodies as particular distinct things. By foregrounding activities and processes facilitated by the permeability of cell membranes, we draw attention to the traffic across the membrane, the influx and efflux, the absorption, recalibration, and response that together shape the biochemical activities within the body's cells and shape the building, dismantling, development, growth, and engagement of that living organism with its social and material habitat. Because of the porosity of cell membranes, an organism that lives in a social and material habitat—as organisms must and do—is unavoidably and ineluctably a biocultural creature.

THREE

Proteins

*In which we learn that constraints in the
conditions for action make a biological process
have direction without intention*

In the previous chapter, I elaborated how the flows of molecules into and out of cells make it impossible to hold onto the presumption that the distinction between the inside and the outside of a cell is a substantive one, that the inside and the outside constitute a substantive distinction. Focusing on the reactions and forms of diffusion conditioned and made possible by such flows, I arrived at the insight that the distinction between the inside and the outside of a cell is a chemical one, a biochemical distinction, a distinction marked by the kinds of chemical activities and processes occurring on one or the other side of the porous cell membrane. In this chapter, I am going to lean on the idea that the distinction between the inside and the outside is a matter of biochemical activity so as to tease apart and disrupt deterministic assumptions about biological causation and to argue that genes are also biocultural phenomena.

It can be disorienting to try to inhabit and fully occupy the notion that the difference between the inside and the outside of a cell is a matter of the chemical processes or activities taking place. It is disorienting because, as Evelyn Fox Keller notes, the mapping of the division between internal and external onto the division between natural biological substances and cultural, environmental phenomena haunts or structures our conceptual vocabulary (2010: 22). Keller notes that even contemporary experts in genetic sciences find themselves recapitulating this mapping in their explanations of research whose findings challenge it. "Everyone knows," she says, that "nature . . . and nurture" are not substantively distinct alternatives (30). "And yet," Keller muses. "And yet, the image of separable ingredients continues to exert a surprisingly strong hold on our imagination, even long after we have learned better" (30). Indeed, I was captured and disoriented by that picture when I first started learning about proteins and genes.

Before my leap into biology, I had, generally speaking, a dietary conception of protein. That is, when I thought of protein, I thought in terms of what we eat: muscles from some unfortunate critter, an element of nuts or dairy goods that makes them an important food source, or a byproduct of a fermentation process through which beans or other vegetables are transformed into highly nutritious edibles like tofu. But proteins do not exist as bare chunks of protein substance located here and there in animal bodies or foods. Rather, as I showed in the previous chapter in my discussion of the pores, channels, and gates that embed in cell membranes to facilitate diffusion, proteins are variously sized molecules that do all kinds of work within and between cells. They perform chemical, structural, communication, and other functions necessary for each and every cell's survival. Nelson and Cox (2008) point out that protein molecules are so central to the ongoing functioning and survival of cells that they form the largest fraction of a cell aside from water (14). Indeed, Garrett and Grisham (2010) claim that if one were to desiccate a cell, 50 percent of its dry weight would be composed of proteins (93).

To accept the idea that proteins constitute a significant chunk of what a cell is and that they are crucial to what it does, I realized that I had to let go of my mind's-eye image of cells as something like water balloons with a paltry few bits floating around in liquid. Instead, I needed to imagine something akin to the bustling, highway-rich city of Los Angeles mapped out in three dimensions. If you can imagine the airways, highways, tramways, railroads, neighborhoods, streets, and thoroughfares in LA, if you can think of all the traffic and transport and housing, factories, offices, public works, and storage, all the creation, movement, destruction, shifting, and recycling of people and goods and information, you begin to have a sense of the structurally complicated, crowded, congested, busy world that is a cell.

In this frenetically active cellular world, thousands of proteins do innumerable tasks. Some join together to form a mesh-like network on both the inside and the outside of the cell membrane, providing infrastructure and shape to the cell. Some proteins stretch the edges of cells, others retract them, and yet others anchor cells together so that they can form a bodily tissue. Some proteins interconnect to stretch across the inner space of the cell, creating pathways or bridges for the transports of chemicals and other proteins around—and in and out of—the cell. Some proteins utilize those pathways and bridges, acting as trucks for the transport for chemicals and other proteins into or out of a nucleus, into or out of cells, throughout the bloodstream to distant parts of the body. And of course, some proteins bypass the bridges and act instead as ferries, shipping their loads through the cell's liquid contents. Other proteins function as channels and gates for those trucks and ferries, permitting or barring entry depending on the kind of chemical seeking to enter or leave the cell. Some proteins bind other proteins together to compose enzymes that make chemicals, break down other chemicals to recycle them for later use, or speed up or slow down that recycling. Yet others store chemicals or package them to be sent elsewhere. All of a cell's proteins together enable the cell to function properly and enable the body to live moment by moment, day by day. Garrett and Grisham (2010) put the point quite succinctly when they say that "proteins fill essentially every biological role," which is to say that they are "the agents of biological function" (120).

In the face of the sheer number and complexity of the many activities undertaken by proteins, one might very well be prompted to ask: how do proteins know what they are supposed to do? How do they know where to go, how to synchronize, what to build or recycle, what and to what extent to regulate and control in cellular function? Given that we do not impute intention or self-conscious deliberation to proteins, how can we account for the ingenuity, the precise coordination, and the efficiency of their actions? From an observer's perspective, each of the many activities undertaken by a protein seems to be done so that the next may also be done; each seems to be undertaken with a view to the next, which is to say purposively or with a reason. But of course, when we reach that point in the train of thought—it seems like they do it with an eye to the process as a whole, in view of, because of, in light of—we know we are being absurd, for molecules do not have reasons and they do not intend what they do.[1]

My bewildered realization that I was inclined to attribute intention or purposiveness to protein molecules in order to understand their very complicated and seemingly punctilious participation in cellular activity was compounded by my initial exposure to the field of genetic science.

When I first started studying genetic activity, I was flabbergasted at the forms of intricacy and regulation—the preparation, production, proofing, editing, and re-proofing—that are integral to the processes through which genes are used to make proteins. In part, my astonishment was due to the digital animation techniques that are used to school undergraduates in the complexities of gene activity: large pastel blobs slide across the projection screen, moving deftly into place to do their work. But my astonishment was not solely a matter of the niceties of pedagogical representation.

As I tried to come to terms with the fact that molecules could work together seemingly purposively in order to make a needed protein, that they could participate in complicated many-stepped processes whose end result was a protein critical for biological functioning, I realized that, in spite of my theoretical training, I had a theological hangover, which is to say that I could not figure out how such processes could be possible without someone, somewhere, knowing what to do. And of course, not being embarrassed enough by my own realization, I said something to this effect to one of my geneticist colleagues—isn't it strange that I have a theological hangover such that attributing intelligence or intention to those molecules seems to be the best way to make sense of their precise coordination in function? And of course, my colleague gave me one of those looks and said something to the effect that he never thinks about it like that. Shoot, I thought: now I feel stupid. But the strangeness and the surprise I felt at that strangeness remained as a spur to thought.

Genes Are Not Überbiological

Let's be unequivocally clear: neither genes nor proteins know what to do. They do not go anywhere nor do anything with intention. The fact that we could think to impute such, even knowing that it is absurd, comes from the legacy of our concepts of biological causation.

Part of my problem in thinking well about the coming-about-of-things in the context of biological organisms is that theorists—having rejected teleological arguments about nature, humankind, and history, through both the rejection of Aristotle's metaphysics and the modern secularization of nature—have also let go of the conceptual tools through which we could account for specificity in biological composition and selectivity and directedness in biological processes. Having rid the natural world, human history, and biology of intrinsic purpose, it seems we have left ourselves equipped with not much more than accident or randomness as a basis for explaining biological activity. Of course, at the level of species, we can turn to theories of evolution for variously deterministic and undeterministic accounts of how species change unfolds (Dawkins 1976; Grosz 2004, 2011; Jablonka and Lamb 2005; Lewontin 2002). But when thinking about biological processes at the level of individual organisms—or even individual cells—how can we account for specificity and direction without making recourse to purpose?

Another part of my problem in thinking well about the coming-about-of-things in the context of biological organisms is the history of scientific efforts to understand genes. Keller (2000) explains that early twentieth-century geneticists were convinced that genes were "the primitive units," the "material elements" of life whose staid, stolid stability enabled them reliably to pass on information from one generation to another about how an organism should develop, grow, function, and behave (18, 47). The conceptualization of genes as "the material basis of life itself" (Haraway 1997: 145) figuratively redrew the boundary distinguishing nature and culture, shifting it to deep within the cell so as to set genes apart from the messy and variable flesh that engaged the world. As Sarah Franklin (2000) glosses this boundary shift, "nature becomes biology becomes genetics" (190). The imagined sequester of genes combined with their purported material fixity to produce a figure of genes as the sine qua non of what is natural in the body. Genes became the überbiological or übernatural matter of life that persists through time and generations.

Keller suggests that the question of *how* an extraordinarily complex adult organism could be built from a mere particulate molecule was initially sidestepped by scientists because the difficulty in answering it was an unwelcome diversion from the efforts to discover *what* genes could do (50). Bracketing such an analysis, scientists simply attributed to genes "a kind of mentality—the ability to plan and delegate" (47). Such an attribution is arguably a holdover from the Aristotelian idea that the paternal contribution to reproduction is the form that gives shape to the maternal matter (Laqueur 1992). Indeed, Clara Pinto-Correia (1997) points out that early images of sperm included a tiny figure of a human crouched in the so-called seed, waiting to grow when deposited in the proper material. Keller contends that such an understanding of the seed was transposed, not without criticism, into what amount to "preformationist" explanations of the causal power of genes (2000: 50). So, not only were genes initially perceived as an überbiological form of matter but they were also conceived as housing or as the origin of some kind of biological force or agency that directs an organism's development and growth.

Of course, the idea that genes are some kind of "master molecule" has undergone excoriating critique. Donna Haraway (1997) is quite blunt in dismissing the notion of the fixed, unwavering materiality of genes, stating that "a gene is not a thing" (142). Instead, she says, "the term gene signifies a node of durable action where many actors . . . meet" (142). Richard Lewontin (2002) similarly unseats genes from a position of biological mastery and highlights instead heterogeneous biochemical activity, explaining that an organism develops as a "consequence of a unique interaction between the genes it carries, the temporal sequence of external environments through which it passes during its life, and random events of molecular interactions within individual cells" (17–18). In fact, in the fast-developing field of epigenetics, scientists work with the realization that in these interactions, some of the biomolecules that are produced while cells and the larger organism respond to stimuli attach to DNA. So attached, they not only affect how and whether genes are used but can also be passed on to further generations of the organism to affect how genes are used in those generations too. These molecules have come to be called epigenetic factors, the prefix "epi-" here designating chemical factors quite literally above and around genes. As I will show in chapter 5, epigenetic factors are so significant in organismic development and growth that, as Keller observes, they put "the very concept of the gene into jeopardy" (2000: 69).

Keller's "and yet" comment noted at the beginning of this chapter suggests that in spite of developments in epigenetics, the initial conceptualization of genes as a kind of überbiological matter continues to haunt the way geneticists talk about genes. More strikingly, as Haraway points out, "belief in the self-sufficiency of genes as 'master molecules,' or as the material basis of life itself . . . not only persists but dominates" in the efforts of nonspecialists to use genetic sciences in their own work (1997: 145). For instance, in recent social science research that seeks to link genes to political behavior, genes are figured as having a determinate or causative effect on growth and behavior (Alford et al. 2005; Fowler and Schrieber 2008). In debates about such research, genes might be portrayed as strongly causing or merely suggestively nudging us to grow or to act like this or like that (Hatemi et al. 2009; Kandler et al. 2012). Whether such portrayals of the activity of genes are accepted or repudiated, the acceptance or repudiation often imagines a far too coherent process, as if the putative action of genes is a cause that governs the entire chain of events that links genes to biological form to behavior. In other words, the parameters of the debate about whether or not genes have a determinate effect on growth and behavior are shaped by an implicit sense that genes are directive (Haraway 1997; Lewontin 2002; Meloni 2014). And while some intrepid social scientists work vigorously to challenge studies that seek to link particular genes with particular behaviors (Beckwith and Morris 2008; Charney 2008; Charney and English 2012), the idea that genes are überbiological master molecules is extraordinarily difficult to dispel.

The aim of this chapter is to scuttle the nonspecialist sense that genes are überbiological master molecules. I shall do so by arguing that biological processes are directional not because genes—or some other unreferenced agent—are in charge but because they are beholden to the conditions of proceeding. In the next sections, I shall explain what such a claim means. But generally, the idea guiding this endeavor is one carried forward from the previous chapter, to wit, if the permeability of a cell membrane means that there is no proper substantive distinction between the inside and the outside of the cell, then there can be no matter inside the cell that is wholly insulated from the inflows from the cellular environs and no causal force proper to or originating from such a(n impossible) substance. If we think about genes and proteins in terms of the constraints on and possibilities for biochemical activity in a cell enclosed by a permeable cell membrane, then we can apprehend the specificity of molecular and cellular activity without having to make recourse to notions of purpose and we can understand how a process can have direction without having to make recourse to notions of intention.

There are two points I want to make in this chapter, then. The first, which I venture to delineate in the next section, is that even though there is no telos, no final cause, intention, or purpose that governs the chain of biological activities that links genes to the production of proteins to the formation of bodies and behavior, those activities are not random or indeterminate. It is not the case that anything goes . . . because only some things and not others are possible at each successive moment (Gunnarsson 2013; Papoulias and Callard 2010). But to say that those biological activities are nonrandom is not the same as to say that they constitute a unified process whose causal arc prescribes and directs the successive individual moments or steps (Massumi 2002). What I will show is that the exactitude or precision of the individual biochemical steps linking genes to proteins to cell function gives direction to the involved biological processes. And that very exactitude and direction does indeed make it seem like there is a process as a whole that is governed by an overarching purpose. However, crucially, the process is not a whole. Instead, rather, there is a succession of distinct conditions for distinct biochemical reactions. Each set of conditions provides the precise occasion for a specific reaction; each moment in the process is the particular condition of possibility for the next. That next is not predetermined or even anticipated. But it is made possible and delimited by the previous one. It is in thinking about the successive production of very precise conditions for reaction that we can begin to think about biological activity in terms of direction without intention.[2]

The second point, which I take up in the last section of the chapter, is that genes are not the starting point of the processes that create the proteins that make many forms of biological activity possible. Even though we can conceive of the processes that link genes, proteins, and cellular function as directed because of the precision and specificity of the conditions for reaction, we cannot understand how and why those processes proceed without what I would call the ecological gesture. As Oyama et al. (2001) insist, in considering genes and the biological processes in which they participate, we must compel ourselves, always, to think ecologically, to think of organisms engaging their habitats over time, habitats whose particular and variable conditions make a specific use of genes and proteins likely rather than not. If we ignore that organismic engagement, with its varied temporal horizon and buffet of provocations, and if we abstract cellular activity from those environmental conditions of an organism's biological possibility, then we may fall back into more reductionist and determinist conceptions of embodiment—bounded, natural, and so forth. In order to account for the extraordinary range and productivity of biological activities particular to cells with permeable cell membranes, we must trace how both the exacting precision of composition and the stimulation of environmental conditions together make the processes linking genes, proteins, and cellular function neither random nor determinate but, in its most literal sense, contingent.

There Is No End

The activity of proteins is directed without a governing intention. There are two considerations that help clarify what I mean by this. The first consideration is protein composition, which both enables and delimits the kinds of activities a protein can do. Because of constraints on the way energy relates to itself when it takes form as matter, the composition of a protein is extremely precise, very exact, astonishingly definite. This precision, which both makes possible and circumscribes the biochemical activities specific to each protein, is what gives directedness to protein activity. The second consideration is the conditions in which a protein subsists. What I mean by conditions here is the array of biochemical substances and molecules bathing and diffusing around a protein in a cell. As biochemical molecules, these substances are what trigger, provoke, or provide the means for a protein's activity. At any particular moment, the particular composition of these biochemical substances makes a protein's activity more or less likely. The volume and variety of biochemicals constituting those conditions make protein activity possible moment by moment. In other words, the variable chemical or molecular conditions that provoke protein activity provide for simultaneous and successive flows of activities. And those flows of activity give spatial, temporal, and biochemical depth to the very precise and delimited individual reactions. If we put the two considerations together, then, we can see that the specificity of each reaction or activity combined with the successive flows of protein activity enable a biological process to have direction-without-intention. I will explain each consideration in turn.

The first consideration is precision. The building blocks of proteins are amino acids. To begin to appreciate what an amino acid is or does, imagine if you will a tripod: three legs connected to a seat or platform, on top of which is affixed a widget, a fancy widget that does useful things. An amino acid is a molecule that has at its center—the seat—a carbon. It is known as the alpha carbon. Suppose we look at an amino acid molecule: the front left leg of the tripod is a small group of molecules called an amino group, and the front right leg of the tripod is a small group of molecules called an acid group. Hence the name amino acid. The third leg, stretching behind the other two, is simply a hydrogen atom, small but terrifically helpful because of its capacity to hydrogen bond. The widget that sits atop the amino acid is called a functional group, a conglomeration of atoms that, because of their chemical composition, affect how that amino acid will interact with other molecules once the amino acid becomes a part of a protein. Just about every one of the twenty amino acids used by our bodies has this structure: the functional groups are what make the amino acids differ from one another.[3]

In order to make a protein, amino acids have to bind to one another—a binding that is undertaken by a complex set of molecular structures inside the cell. And, strikingly, when amino acids are brought into proximity to bind to one another, they always bind in a very particular fashion: the amino group of one alpha carbon will interact with—which is to say that it will bond or join with—the acid group of another alpha carbon. The resulting bond is called a peptide bond. And now that the former amino acids are bound, they are together known as a peptide (and many strung together are known as polypeptides).

What is important to understand here is that the addition of further amino acids to our initial count of two is directional. What I mean by this is that the other amino acids add only to one end of the string—and never to the other or to both ends. As amino acids are serially added to a string by the complex set of molecular structures that bring the amino acids into proximity, it is always, always the case that the amino group of the newcomer binds to the acid group of the string. As a consequence, every polypeptide has an amino end that counts as its beginning and an acid end that counts as its termination point. (Think here of the linked tripods: a free left leg on one end, a free right leg on another.) These ends are known respectively as the N-terminus (N for the nitrogen in the amino group) and the C-terminus (C for the carbon-oxygen atoms in the acid group). This is helpful to know because when scientists identify and talk to one another about particular amino acid residues in a protein, they number them with the number 1 starting at the N-terminus: we all start in the same spot and count to pinpoint a specific amino acid. It is also helpful to know because polypeptides begin folding into functional proteins from the N-terminus.

Note, here, that the very particular shape of amino acid molecules and the very particular manner in which they join together delimit how a polypeptide is composed. The precision in the composition of the amino acids delimits the process of polypeptide formation. Precision provides the constraints that make the compositional process have direction. The delimiting and directing effects of precision in composition are critical for understanding proteins.

At this point, we have in our imagination a stringy polypeptide with a definite beginning and a definite end. We do not yet have a protein. A protein is formed when the string folds—and folds in a very particular fashion. Imagine, in your mind's eye, suspending a fairly long piece of string above your desk. Slowly lower your imaginary hand to enable the string to gather and bunch into a clump on the desktop. If you then use your hand to smoosh the string together a little, what you end up with is a variably dense tangle with no knots. Although ungainly as a figure, this image gives us a useful snapshot of what a protein looks like. A protein is composed of a polypeptide that bends, twists, folds, and turns—but in much less random and indeed much more precise ways than the imagined string on your desk. In fact, depending on what those amino acid molecules are and the order in which they are strung together, the string will fold, gather, bunch, and clump in ways that enable the resultant protein to do very particular things.

The folding of an actual string of polypeptides (as opposed to the folding of our imagined string) occurs in a very specific fashion: the order or sequence of amino acids in a string of peptides makes it possible for the polypeptide to fold in one or another particular way. There are two dominant forms of folding. In some sequences of polypeptides, bits and pieces of the connected polypeptides interact via their electrical charges in such a way as to result in a twist or spiral along the peptide bond. As successive peptides twist, the polypeptide string comes to resemble one of those (now) old-fashioned telephone wires or a slinky: a spiraling coil known as an alpha helix. Another predominant form of folding is a zigzag crimping called a beta strand. Here, the interaction of various elements in the polypeptide cause the string to fold back and forth, bit by bit, akin to the folding of paper to make a paper fan. If a zigzagging beta strand is quite long, it can wind back and forth so that all the ups and all the downs are aligned to create a blanket-like beta sheet. Then, depending on the riot of electron activity constraining and facilitating these bends, kinks, and swirls, as Lewontin (2002) explains, the helices, strands, and sheets fold on themselves to create a protein whose pockets and protuberances are the sites of biochemical activity (74). It is exactly because of the sequence of the amino acids in its composition and the changing vibrational constraints throughout the molecule that a polypeptide folds in just such a way as to produce sites of specific biochemical activity.[4]

So, once a peptide has folded, it becomes a functional protein. What its function is—what it is capable of doing—depends on where the functional groups on the amino acids that compose the protein lie (the widgets!) and what those functional groups are able to do in relation to other proteins or biochemical molecules. Thus we get one of the key principles guiding scientists' understanding of proteins. The specific sequence of amino acids gives rise to the particular pattern of folding and structure that gives the protein its specific function: sequence → structure → function. Exactly how does protein structure both circumscribe and direct protein function?

When a polypeptide folds to create a protein, the juxtaposition and attachment of the amino acid residues to one another compose a variously lumpy, bumpy mass. The topography of that mass is critical to each protein's function: each valley and crevasse, each butte and rise, creates a pocket or an appendage where a very particular amino acid with a very particular widget (aka chemical shape or electrical charge) can be sequestered or made available for interaction with another molecule. The charge and shape of that topography are such that if a molecule with a complementary chemical shape or electrical charge tumbles by, it is likely to slip into place, to attach and to react in a manner specific to that joining.

We could say, then, that a protein folds in such a way that its shape constitutes the conditions for its reactive interaction with other molecules. A reactive interaction occurs when a molecule passes by that "fits" the protein's reactive part in the way a hand fits into a glove, a key fits into a lock, or a puzzle piece fits into a particular space. Lodish et al. (2008) call this draw and fit "molecular complementarity" (32, 39, 78); Garrett and Grisham (2010) call it "molecular recognition through structural complementarity" (14, 122–23). What both descriptions tell us is that the precision in the activity of proteins comes from the fact that the shape of each specific protein mass makes it well suited to some very particular tasks and ill suited to many others. The specificity of each protein's shape is both enabling and limiting; it makes particular interactions possible and restricts possible interactions to those particular ones. Precision is a constraint. And precision is a condition of direction in biological activity.

The second consideration. What is important to point out here is that the reactive interactions made possible by the exact sequence and precise folding of a polypeptide into a protein can take place only with the availability of the molecules that circulate around the protein in the cell. The precision is a condition of directed protein activity, but not the only one. Just as a left-handed glove needs a left and not a right hand, a specific skeleton key hole requires a specific skeleton key, and a knobbly gap in a puzzle can be filled only by its corresponding puzzle piece, so the reactive interactions of proteins can occur only when a protein is in the presence of molecules that are structurally complementary to particular reactive areas of that protein. In other words, the volume and kind of biochemical substances surrounding a protein are a condition for protein activity to proceed.

Among the very many chemicals and molecules that roil and swirl in and around cells in accordance with the forces of diffusion and biochemical reaction, there will be some that complement and can interact with particular proteins. They have to be in the right place at the right time, facing in the right direction, nudging the protein in just such a manner as to enable the interaction to take place. Put like this, though, the requirements of complementarity make such a reaction seem terribly unlikely: what are the odds?

At the molecular level, each step in each sequence of the protein activity that makes cells function is improbable but not impossible. However, the huge volume of molecules roiling, diffusing, and shifting concentrations as they swirl by any given protein changes the disposition of a reaction from seemingly random and unlikely to really likely and rather predictable. For many living bodies, the infusion of molecules across the porous cell membrane from the cellular or organismic environment provides such a preponderance of converging possibilities that particular protein activities—and the succession of those activities—are fairly regular. The volume of molecules available for a protein's activity rises and falls depending on how the cell or organism responds to the demands the environment makes on it. What we can see here, then, is that the flow of protein activity is dependent on whether the influx of substances into the cell is at the concentration or volume that makes such activity probable. In other words, it depends on the body's situation.

The precision of protein composition and the conditions of activity provided by a reliable volume of reactive molecules together make protein activity seem to be targeted, purposive, intentional: directed to an end. This seeming is all the more pressing when we consider that most of the activity undertaken by proteins in a cell entails multistep processes involving a goodly number of proteins: it seems like they are working together. But what our investigation of proteins suggests is that each interaction or reaction creates the conditions for the next without anticipating it: one step changes the shape of the protein or amends the structure of a biochemical molecule so that it becomes the condition for another interaction with another protein or another molecule—each of which is also the condition for that particular interaction. It seems like there is a seamless directed process. But instead there is precision and condition, direction without intention, direction with no end.

There Is No Beginning

As I showed in the case of the pores, channels, and gates that are embedded in cell membranes, both the influx and efflux of chemical molecules across the permeable cell membrane and the variation in concentrations of different kinds of molecules provided by biochemical reactions in the cell provoke protein activity. What this means is that proteins are responsive agents in rather than permanent features of our cells. As Kristensen et al. (2013) observe, rather than floating around, precomposed, waiting for an appropriate task, proteins develop in response to a biochemical need in the cell. And Nelson and Cox (2008) note, once composed, proteins last only for a relatively short while—anywhere from a couple of minutes to just over ten days (572).[5] Indeed, it is quite incredible to imagine that something as complicated and dense as the city of Los Angeles—and something as critical to cell survival as all those proteins—could be dismantled, recycled, and rebuilt on a minute-by-minute, hour-by-hour, day-by-day basis. Of course, a living organism is constantly responding to its environment, and since proteins can take some time to build, at any given moment there will be ongoing patterns of protein production as well as proteins at work so that the cell and the organism are ready for whatever stimuli or need they encounter. The point of remarking on the responsiveness of proteins is that it is because proteins are dismantled, recycled, and rebuilt so very frequently that genes are important.

For while we often think about genes in terms of reproduction through generations, which is to say in terms of evolution and inheritance (and, in human politics, population management and eugenics), they are important in a daily, minute-by-minute way because they constitute the recipe for all the proteins that keep organisms alive. As I showed in chapter 2, the permeability of cell membranes enables organisms to perceive and respond to the stimuli that constitute their habitats. As an organism encounters and engages with its environment, its cells respond to the provocations with various chemical releases and hormonal cascades. These bathe the cells, infiltrate them, and stimulate the work of proteins that either produce or release other chemical molecules or that facilitate chemical reactions that enable the cell to recalibrate and persist in its complex of biochemical activity. If there are not enough of those proteins for the cell to function well as it responds, then more proteins are made using genes as their template.

It is in the process by which genes are used to make proteins that we see that there is no beginning. What I mean in saying this is that genes are not the origin of their own activity. They direct nothing. They cause nothing. As I will show, genes constitute a fantastically well-preserved, highly modified, tremendously precise recipe for making proteins. But the need for a protein's activity must be felt by a cell, the ingredients must be supplied, and the instructions for how to use those ingredients must be accessed, read, and followed for that gene recipe to result in the composition of a protein. To put the point differently, we rarely ever think of a recipe as "making" that for which it is the recipe; even when we speak very colloquially of a recipe making two dozen somethings, implied in what we know about recipes is that the activity is undertaken by someone other than the recipe itself. Similarly, we need to shift ourselves into a framework in which it sounds silly to say that a gene does something as if it is a cause, action, or agent. Instead, it is the referential basis for activity undertaken by others, by biomolecules and by proteins under the needful prompting of cellular perception.

According to Nick Lane (2010), scientists speculate that in primordial biological history, proteins were made by conglomerations of molecules called ribonucleic acids, also known as RNA (from RiboNucleic Acid). A ribonucleic acid is a large molecule composed of a string of small ribonucleotides, each of which has a carbon ring with some appendages, including a ribose or ringed sugar. These ribonucleotides string together in much the same way as amino acids, in a long string that folds into clumpy lumps that do all manner of chemical things. However, rather than twenty possible types of molecules to add to the string, as in the case of amino acids, for RNAs there are only four. This limits the kinds of chemical reactions they can foment, but they are reactive enough that they can capture specific amino acids and foster their binding to create proteins and engage in other regulatory functions in cells.

Lane (2010) explains that the very ribose sugar ring that made RNA so very reactive and effective also made it unstable. The ribose sugar ring includes oxygen-hydrogen appendages that are incredibly reactive. On the one hand this is terrifically useful, because it means that the RNA molecules can readily participate in a variety of reactions. However, and on the other hand, the reactivity of those oxygen-hydrogen appendages means that they can easily be a target for reactions—ripped off or compounded—and thus a place of instability in the molecules themselves. As a consequence of this instability, RNA is unreliable as a source of memory for how to construct proteins: any long-standing string of RNA that might be used as a kind of template to build a protein could be corrupted or transformed—and the model thereby lost.

As Lane (2010) tells the tale, over time, biological processes developed whereby one of the instability-inducing oxygen-hydrogen appendages was removed from the ribonucleic acids that form the memory for constructing proteins. Missing one of the oxygen groups—de-oxygenized, we might say—it is known as DE-oxy-ribonucleic acid, which is abbreviated as DNA. Because it is missing that oxygen group, it is much more stable over time, much less liable to be corrupted either by the addition of something to that oxygen or by the ripping off of that oxygen by another molecule. Over time, then, the memory of how to make proteins and the actual protein-making were divided between the RNA and the DNA, with the DNA providing a molecular memory of the proteins necessary to construct and maintain cells and the RNA providing the tools to make that molecular memory into real proteins.

DNA, the molecular memory of the specific strings of amino acids necessary to create specific proteins, is housed in every cell's nucleus. And like a rare manuscript in a rare manuscripts library, it never leaves. And what comes into the nucleus is strictly regulated via gates so that damage to or corruption of that DNA can be minimized.

The fact that the integrity of DNA is critical to the production of functioning proteins can be seen both in its structure and in the precision with which it is used.

Structure first. Except for a couple of adjustments, DNA is very similar to RNA: there are four different nucleotides, guanine, cytosine, thymine, adenosine, which are generally referred to by their initial letters (G, C, T, and A). The nucleotides are joined together in a string that is paired with another string whose constituent pieces are a reciprocal mirror image of those in the first: G always mirrors with C; T always mirrors with A. Together, these strings compose a ladder or a so-called double helix that twists and folds on itself to create a dense mass. The mirroring in the structure of the DNA provides the conditions for precision when the DNA needs to be copied so that another cell can be made.

During copying, a special RNA-protein assemblage unfolds the dense mass and then unzips the double helix so that each strand is separate. Each strand is then appended with a replicator (again made of RNA and proteins) that, through molecular complementarity, recognizes which particular DNA nucleotide is in an internal pocket and draws in from the volume of molecules swirling around the nucleus its reciprocal match and helps them bind. So, to illustrate with an example—without sounding too much like Dr. Seuss—imagine, if you will, a ladder composed of string 1 and string 2.

TATAAAGCTGT (string 1)
ATATTTCGACA (string 2)

When separated, the T on string 1 will be matched with an A pulled from the volume of biochemicals floating around in the nucleus; and the A on string 2 will be matched with a T.

If this matching is repeated down the two separated strands, eventually there will be two entire complete double helices that are an exact copy of the original. These copies will be apportioned as the cell divides into two.

The structure of DNA enables the copying or replication processes to be very precise. Indeed, following behind the replicator assemblage that reads the string and binds nucleotides to it is a molecular assemblage that functions as an editor detecting whether a proper match has been made: if not, then the offending interloper is cut out and another appropriate nucleotide is spliced in. Such exactitude ensures that the sequence of the nucleotides that tells how to make proteins is not inadvertently messed up in the process of copying.

Now, precision in use. The relationship between DNA and proteins is quite complicated, for the genes are not the origin of their own activity, their own being-used. And here, again, the analogy to a rare book manuscript is perhaps helpful. When the proteins in a cell respond to a cellular or organismic demand—a demand that manifests at the molecular level as biochemical activity—that demand results in chemical signals entering the nucleus to initiate the production of the particular proteins necessary to facilitate a particular response. In other words, the trigger for genetic activity comes from the cell or the organism as a whole as it responds to its environment. But the trigger is a much more complicated and highly regulated process than just the appearance of a chemical sign. To understand how so, we must recognize that DNA is not the same thing as genes.

Genes are the particular strings of DNA that code for proteins: there are around 25,000 genes in the human genome. However, genes account for only around 2 percent of the DNA in the genome. In large part because of the governing understanding of biological causation, the other 98 percent of DNA in the human genome was, until recently, considered "junk" DNA, relics from the past, sloppy leftovers and such that were no longer needed (Keller 2000: 59). However, Keller notes that scientists now increasingly realize that much of this "junk" could actually be crucial to the process of making the thousands upon thousands of proteins that our bodies need to survive (59). Some of the so-called junk DNA codes for the RNAs that carry out so much work in the making of proteins and in facilitating chemical reactions in the cell. And some of that junk DNA serves regulatory functions that promote, facilitate, delay, accelerate, and specify how the genetic material should be used. Which is to say that a lot of the noncoding DNA is mobilized to ensure precision in the use to which the genes are put. To understand how, we need to go into a bit more detail.

We often hear talk of DNA being a code. It is a code in the sense that the individual DNA nucleotides do not singly point to individual amino acids. Such a one-to-one correspondence is actually not possible, because there are twenty different amino acids and only four DNA nucleotides. So, instead of being a one-to-one correspondence, there is a three-to-one correspondence. Each group of three nucleotides in a strand of DNA composing a gene stands for or "codes" for one particular amino acid. Here is an illustrative example with our own English alphabet. If we were to see a string of DNA composing a gene that read THEFATCATSAT, we could apply the three-to-one reading frame to see the following sets of three: THE FAT CAT SAT. Each of these individual three-letter words would code for a specific amino acid, which means that this string would be a gene coding for four amino acids.

There are two really important things to recognize about this kind of coding. The first is the importance of the integrity of the DNA strand—especially in copying or replication. For if the reading frame were shifted—let's say that the *E* of THE was mistakenly dropped from the sequence and then all the letters shifted to make three-letter words in the reading frame, the first three amino acids would be completely different, and the last would be an incomplete dud: THF ATC ATS AT. If an additional "letter" is added, or if two letters were switched, similar muddles would follow. In either case, what results is a set of instructions for a protein that are either rubbish—unintelligible—or slightly erroneous. For example, if the *C* on CAT were to switch to an *F*, the word would still read as a word even if the sentence was not completely sensible. The protein built from such a mistake might be mostly right, mostly functional, and therefore might not contribute to the death of the cell. But then again, if the amino acid that was supposed to be there were an absolutely critical part of the overall protein—in terms of its structure or function—then the protein might be unusable. The integrity of the DNA strand is critical to the specificity of composition that enables proteins to fold and to act in specific ways.

The second important thing about this coding is that the sequence of genes need not code for just one sequence of amino acids—just one protein. As mentioned, our imagined section of DNA that we call a gene may look something like this: THEFATCATSAT, which would give us a particular protein composed of four amino acids. Yet, if we include in our example the various pieces of not-actually-junk DNA interspersed in and surrounding that sequence, it might look like MMMMTHEMMMFATMMMMMCATMMMMSATMMMM, with all the *M*s, as regulatory DNA, signaling to the RNA molecules to cut or splice the sequence under different circumstances so as to produce other possible proteins: THE CAT SAT, THE FAT CAT, CAT CAT CAT, THE FAT THE FAT THE FAT. And which of those possibilities should be brought into play is signaled by other molecules in the cell that, in response to chemical signals regarding recurring conditions or new ones encountered by the organism, attach to Ms or to sections of the DNA that the Ms regulate so as to direct and effect the use of genes to make the specific needed proteins. We can see here, then, that the Ms—the regulatory genes—make possible the many different kinds of cells we see in a living organism (Klug et al. 2009: 84). That is, even though all cells in a given living organism hold roughly the same DNA, the regulatory genes, responding to biochemical molecules emitted by neighboring or nearby cells and tissues, constrain and facilitate the production of the proteins that will make "liver cells, skin cells, and kidney cells, look different, behave differently, and function differently" (Jablonka and Lamb 2005: 113).

Now, before moving on to a neat summation of how genes are used to make proteins, I want to pause to focus on the particular activity and provenance of the molecules and chemicals that trigger the use of genes to make proteins. Some of these molecules actually attach to the string of DNA as molecular equivalents of sticky notes.[6] They accumulate over time as a result of habitual responses of the cell and organism to various stimuli. And, strikingly, they can be passed on from one generation to the next and to the next. As mentioned earlier in the chapter, these sticky notes are known as epigenetic markers because they have an effect on which genes will or will not be used, which proteins will or will not be made, even though they are not themselves genes. Epigenetic markers may do any of several things: make the transcription of the spot to which they are attached easier, make the molecular transcription machinery repeat the transcription a number of times, prevent transcription altogether, or work in combination with regulatory genes and other not-actually-junk DNA to prescribe whatever pattern of transcription might have emerged as useful from the longer-term response of the cell or organism to such needs.

And, strikingly, sometimes some among these epigenetic markers and chemical and regulatory molecules take a DNA transcript and do something that is called reverse-transcription, in which the transcript is used to make extra copies of that section of DNA that then get reinserted into the DNA sequence. As you might infer, this kind of reverse-transcription changes the recipe for the protein, changes the gene sequences in that particular cell. This means that an individual organism can be composed of cells that have different DNA—a phenomenon known as somatic mosaicism (as in "a mosaic of cells with different DNA") (Charney 2012; Lupski 2013).

There is a related phenomenon called chimerism, which involves the insertion of DNA from *other* organisms into the DNA of cells—which is thought possibly to occur through gestation, pregnancy, blood transfusion, organ transplantation, or sexual activity (O'Huallachain et al. 2012; Yan et al. 2005). According to Hall (1998), mosaicism and chimerism were initially thought to be linked to diseases like cancer and neurodegenerative disorders. However, Lupski (2013) suggests that such phenomena might actually "play a role in normal biological function" (359). And, more surprisingly, Chan et al. (2012) speculate that these phenomena might play a prophylactic role, boosting the immune system, preventing the onset of diseases like Alzheimer's, and possibly having an important role to play in evolution.

So, finally: to make a protein. When an organism undergoes some kind of perceptual stimulation, or when the organism's body or environment provokes a response, or when an organism develops, grows, and daily refurnishes its vast collection of proteins, the kind and the concentration of molecules crossing the cell membrane change. This change sets in motion a chain of chemical reactions that result in masses of protein and RNA molecules coassembling to become transcribers of genes. Which ones coassemble and where on the strand of DNA they dock will depend on the exact nature of the chemical changes and the related reactions occurring in the cell: what kind of protein is needed to address the chemical provocation? If the regulatory sticky-note molecules attached to the DNA strand allow transcription, the RNA-protein assemblage will unravel that portion of the tangled mess and copy the DNA according to the molecular conditions, promotions, and circumscriptions. The transcripts they produce are an exact copy of the DNA sequence. The transcripts are edited, spliced, and refined by other RNA-protein assemblages that, along with other various chemical signals, specify which particular pieces of the sequence should be used and which can be discarded. The edited transcript is then ferried out of the nucleus by other proteins, through gates (which are also proteins), into the larger internal cell space. These refined transcripts are then shuttled to protein factories called ribosomes that read the transcripts and translate each group of three into the directions for assembling amino acids into proteins. As each group of three comes to sit in a reactive pocket in the ribosome, the corresponding amino acid is grabbed as it floats by and appended to the growing string of peptides . . . which, as I have shown, then folds to create a functional protein.[7]

The Biocultural Organism Lives

This chapter has argued that we should conceive of genes as biocultural. Genes are not static forms of molecular matter. Rather, they are dynamically maintained and frequently transformed by an expansive set of molecular RNA-protein assemblages within the cell. Keller puts this point quite nicely when she observes that "the stability of gene structure thus appears not as a starting point but as an end-product—as the result of a highly orchestrated dynamic process" (2000: 31). Furthermore, genes are not walled in or encased in a protective covering that shields them from the incursions of chemicals (or viruses). Quite the contrary: not only do a wide range of molecular formations touch, cut, edit, amend, or append themselves to DNA but also the constant influx and efflux of chemicals across the cell membrane precipitate and define the trajectory of protein production. The signaling molecules that time and experience append to DNA sequences, the promoters, the docking recruiters and enablers, the transcribers, editors, refiners, and transporters, the translators, proofers, and splicers: these RNA-protein assemblages all do their work with DNA in short- and long-term response to chemical signals that the habitat-engaging body generates. All of this busyness in the vicinity of DNA puts a considerable wrench into the idea of a sequestered, protected individual genome. And finally, genes are not themselves active; they do not direct other molecules, nor is some action or force on their part the origin of their use in the production of a protein. Genes are a well-preserved, highly modified reference point in the ongoing efforts of a living organism to survive and thrive in its habitat. Indeed, the chemical molecules that are generated by an organism and that precipitate or provoke the use of DNA concern not genes or DNA per se but rather the quantity and type of proteins that an organism's cells need in order to function as the organism encounters its world.

There is no überbiological matter in a cell, in a body. What this means is that far from being isolated, constitutionally durable, and materially infallible things, genes—and proteins and cells and living bodies—are biocultural phenomena. Because they are biocultural, we must always think of them in terms of a living organism that responds to its needs over time, through its life history, in a specific context, and in response to events (Jablonka and Lamb 2006; Lewontin 2002). In other words, this chapter begins to demonstrate—as the next will continue to explain—that when we think about biological processes and activities, we cannot but consider an organism as it engages, perceives, responds to, reaches toward, and imbibes its habitat in its efforts to persist in living.

FOUR

Oxygen

*In which we reaffirm that no living organism
can survive deprived of an environment*

In the last chapter, we mobilized the idea that cell membranes are constitutively and selectively permeable to underscore that no feature of a cell is isolated from the continuous influx and efflux of molecules across those membranes. In this sense, even genes are biocultural; the cellular activity of which they are a part is bound up with the ways that an organism responds to the provocations of its habitat as well as to patterns and interruptions within its own biological processes. In this chapter, I will consider what it means to conceive of that influx and efflux as a movement bound up with changes in energy rather than a movement of substances. One of the insights this chapter develops is that if we conceive of the traffic of molecules in terms of the movement of substances, then we also figure the influx and efflux as an addition or subtraction of substances. But if we conceive of that traffic in terms of shifts in energy that is constrained in the way it relates to itself, then we must figure the influx and efflux as effecting transitions in energy. The larger point of this chapter is that if we conceive of the influx and efflux in terms of energy-in-transition, then we can discern more sharply how a living body relies on and is sustained by that traffic. It is not just that the permeability of cell membranes effects a continuous influx and efflux of molecules but also that there must be that influx and efflux. Living organisms do not just happen to exist in an environment that filters into and out of their cells. Instead, they are critically dependent on an environment filtering into and out of their cells. One of the aims of this chapter, then, is to highlight the importance of thinking about what Lynda Birke calls the "whole organism" (Asberg and Birke 2010: 417) and what Kültz et al. (2013) describe as "the integrated organism"— even when working and thinking about creaturely embodiment at the molecular and cellular level. (See also Roberts 2003.)

When I first started to reflect on the notion that energy is at play in cellular activities, I could track the molecular patterns conceptually and work roughly with their logic. But when the course material demanded that we integrate what Myra Hird (2009b) calls the "big like us" scale of gross organisms that eat and breathe, my grasp of the concepts ran up against my casual everyday capture in a substance-oriented framework.

For, while I was quite able to acknowledge that eating a sugary morsel can lead to hyperenergetic behavior, and while I knew that the calories by which we tabulate food intake are a measure of energy, I had not really absorbed what it meant to conceive of food in terms of energy. When I learned what happens to the food that we ingest, I realized that I had held on to a notion of food as something akin to building blocks that, when ingested, add to and become part of a bigger, more ample lump of substance. But when living bodies break down food—let's say, a sugary morsel— it is not that the cells need or use the carbon, hydrogen, or oxygen atoms of which those carbohydrates are composed. Rather what is crucial is the process of ripping those atoms apart from one another and all the electrical transitions and molecular rearrangements that are a constitutive part of that dismantling process. In other words, what is "used" by cells in the body is the process of ripping apart and the transitions in energy that such ripping makes possible. These transitions in energy are the conditions for further transitions in energy that take the form of diffusion or chemical reactions. Indeed, once the reorganization and ripping apart has taken place, the individual atoms that had composed the food end up being something akin to waste products that the cell must either dispose of or recycle for different purposes and reactions. For instance, the carbon atoms that remain from the dismantling of sugary carbohydrates are the carbons in the carbon dioxide molecules that we exhale when we breathe.

This was a wonderfully disorienting desubstantialization of food that gave me great pause for thought. It was all the more so when I learned that the oxygen we breathe in with air is used by cells in the disposal end of that ripping-and-disposal process.

What was unsettling here was not just my sense that eating and breathing were two different things—as if the solid or liquid substance of food should be involved in completely different processes in the body than the gaseous substance of oxygen because we take them in differently and because, relatedly, food is solid or liquid rather than gas. What was equally unsettling was that, in my unschooled common sense, I had tended to think that oxygen feeds something; that it is important because it enriches or adds to something. We all know that oxygen is extremely important for our survival. We need oxygen—and one feels that need desperately when one's ability to suck it in as one breathes is compromised, either because of sickness, because there is an obstacle in one's airways, or because one is exercising beyond one's comfort zone. But beyond a strong and clear conviction that we need oxygen, things tend to become fuzzier. We breathe oxygen into our lungs; from our lungs it enters the bloodstream, where it makes our blood brighter, redder, and makes us feel better. But as the vagueness of "makes us feel better" suggests, exactly what oxygen is doing to make us feel better is not part of our popular self-knowledge. What is oxygen doing in there? How is it "used up" such that we need to add more—such that we need to keep breathing?

As noted, oxygen is critical not in what it adds to a cell or to an organism but in what it takes away. When we take in oxygen through breathing, the oxygen participates in a process called cellular respiration. Cellular respiration is a metabolic activity, which is to say it involves the transformation of food into chemical energy that can be used by the cell in its various and sundry jobs. Breathing enables the body to take in the oxygen that will enable cellular respiration to take place. What does oxygen do that is so central to cellular respiration? Oxygen is a trash carrier at the tail end of the processes through which a cell breaks down sugars to produce the energy needed to fuel difficult reactions. Oxygen arrives on the scene at the last crucial stage in the processes of cellular respiration; in that last stage, it captures and removes electrons and hydrogen ions so that the processes of making energy molecules can continue.

In this context, then, oxygen is not a compositional but rather a transient molecule. It is not that oxygen needs to be there—as in, to add itself to the cell—but rather that it needs to have been there, to have passed through. It arrives in order to leave. We could say that its modality is that of passage: oxygen has to appear and leave, arrive and pass on. In fact, it is only in its passage that oxygen fulfills its crucial function.

My tendency to think that oxygen makes what it is added to better and more richly composed revealed to me that I was working within a conceptual framework in which bodies are substances and the process of adding substances to bodies is a process of composition. But if oxygen is a transient molecule, then the conceptual framework of addition and composition does not capture adequately the role it plays in living bodies. The challenge of this chapter is to trace the contours of the conceptual shifts attendant to thinking of oxygen in terms of the transitions in energy it facilitates.

Energy in Transition

The idea that eating and breathing are additive processes is cognate with the notion that a cell membrane should separate the substances on the inside from the substances on the outside—in such a framework, ingestion subtracts material stuff from the world beyond the body and thereby adds material stuff, like building blocks, to a body. But here, I am elaborating the possibility that the distinction between the inside and the outside of a cell or a body is not properly a substantive one but rather one centered on biochemical activity. In this framework, we should see food and air as taken into a living body for the purpose of facilitating the transitions in energy that cells use to provoke reactions and the transformation of molecules from one state to another. Indeed, if we cleave to this idea of energy-in-transition, if we remember that the biochemical reactions that are the basis of biological activity are essentially changes or transitions in energy as molecules bond or break apart, then we can capture conceptually the ways that the ingestion of food and the aspiration of air are essential elements of always ongoing activities of composing and decomposing.

In pushing the idea that we conceive of living bodies in terms of energy-in-transition rather than in terms of stuff-built-or-composed, I do not mean to imply that there is no stuff of which the body is composed. Clearly, this is not the case. Rather, in focusing on energy-in-transition, I want to mitigate the possibility that we might slip into the habit of thinking of composition in terms of a finished product—a having-been-composed, a static substance, a stable status—rather than thinking about composition as the activity of composing and decomposing.

However, as I began thinking about oxygen in terms of the transitions in energy that it makes possible, another puzzle arose concerning the vital importance of oxygen for living bodies. When I realized that oxygen is crucial not in adhesion or supplementation but rather in its transience, my sense of embodiment—of what embodiment is—was seriously destabilized. To conceive of oxygen from within a framework of transition rather than addition seemed to be to render the body almost incomprehensibly loosey-goosey, insubstantial, ephemeral, or unreliable: less *there*. It seemed to me that to think about oxygen in terms of its transience was to undermine the "itness" of the oxygen as well as the cell and the body more generally. And, strangely enough, it seemed as if oxygen had been rendered less important somehow.

But of course, oxygen is important. In the gaseous form in which we breathe it in, it is crucial, a without-which-we-cannot-survive kind of molecule. And indeed, the use of it constantly, moment by moment, is one of the defining distinctions between two major forms of life: plants and creatures.

My sense of unsettled disappointment in oxygen's role in cellular respiration—even in the face of the recognition that it is crucial for an entire form of life—became emblematic for me of the ways the conceptualization of living bodies in terms of energy-in-transition demands that we recalibrate the criteria according to which we assess or measure what is important in living organisms. My sense of disappointment came from a presumption that the importance of oxygen must rest on its particular activity in a process, and more specifically that it must rest on oxygen having some kind of outsized causal efficaciousness compared with other molecules. I presumed, that is, that oxygen is critical because it is the figurative engine of the processes in which it participates. In chapter 3, however, I showed that no one molecule can be considered a "master molecule" driving or controlling a biochemical process. What this means is that the importance of oxygen is not to be measured by the size or extent of its role. In the processes of cellular respiration, oxygen is not a principal player in the sense of being a primary agent appearing in every scene. Oxygen actually has only a cameo role at the very tail end of the processes of cellular respiration; its very tardiness in appearing in the narrative of the processes in which it is such a crucial player only underscores the peculiar nature of its importance. Indeed (a warning) a reader of the processes in which oxygen participates might become puzzled and somewhat frustrated by the seeming deferral of oxygen's appearance in the narrative: where is it?

As I showed in chapter 3, biochemical processes unfold via step-by-step reactions, with each moment constituting both the limiting and enabling condition of the next. Accordingly, as I shall show in the account laid out in the next two sections, every step in the processes of cellular respiration is important because the transition each step effects is the condition of the next. Among all those steps, the arrival of oxygen in a cell to extract the trash it must extract—this arrival is one that depends critically on the relationship between the body entire and its environment. In other words, oxygen gains its importance not for an outstanding or premier role, not for a particularly extraordinary contribution to an otherwise humdrum process, but for its vulnerability to disruption, for its utter dependence on the organism's interface with an oxygen-rich habitat.

Whereas many of the steps in cellular respiration direct our attention to cells, molecules, and their busy activities, oxygen demands that we also, at the same time, attend to the organism that is composed of cells, the organism that breathes, the organism whose breathing has systemic effects. For if breathing fails, if the organism does not take oxygen in, then cellular respiration halts not just in one cell, or in two or three, but in all cells, in the organism as a whole. Oxygen highlights the fact that the porous cells that compose bodies compose an organism. Which is to say that oxygen recalls us to the fact that in addition to being a living system composed of cells, an organism is an organism.

But this is not yet the whole point. The itinerant nature of oxygen underscores a further entailment of conceiving of bodies as energy-in-transition, as composites of chemical processes, as composing and decomposing. For organisms, composing themselves in the manner in which they compose, are crucially dependent on the traffic and transit of chemicals and molecules back and forth between their cells and their habitat. And just as energy cannot be created from itself but instead can only be transferred, so a porous living organism constituted through transitions in energy cannot be self-sustaining in any sense of the term. The passage of oxygen via the porosity of cell membranes brings into focus not only that the interchange and transit of molecules in and out of cells is possible but also that it is critical for the living organism's survival. Indeed, without that interchange, without that multifarious influx and efflux, the biochemical processes that sustain organismic life would not be possible at all.

When Diffusion Is Not Enough

To trace the peculiar work of oxygen in a body conceived in terms of energy-in-transition, we need to hold before our minds two interrelated points. The first point concerns the way that, when energy takes form as matter through its constrained self-relation, it has a tendency to shift into the least tense state possible. As I showed in chapter 2, many of the biochemical activities in cells occur in such a manner as to effect a decrease in energetic tension. This diminution of energetic tension can take place through the dispersion or diffusion of a concentration of a particular kind of molecule or substance. It can also take place through a bond-making chemical reaction that changes the configuration of the molecules involved.

The second point concerns the fact that diffusion is not enough. Just as one can run down hill only for so long before one is compelled to run up again, so the energy-changing reactions that constitute the bulk of cellular activity can diminish energy only so much before they must rise so as to diminish again. Accordingly, many of the body's activities entail ratcheting up the levels of energy in order subsequently to be able to let them diminish: run it down, ramp it up, run it down. Transition. The ingestion of food and the breathing in of oxygen create the conditions for that ratcheting up of energetic tension so that biochemically necessary tension-reducing reactions can take place.

As I have shown, the activities that occur within and between cells are all essentially chemical reactions, interactions between molecules that form bonds, disintegrate, transform, and then interact with other molecules in different ways. Many of these reactions both require and produce shifting concentrations of molecules, which is to say that they rely on and provoke processes of diffusion. And the selectively permeable cell membrane that encloses each cell facilitates and enhances the processes of diffusion. In many cases, however, reactions facilitated and provoked by the processes of diffusion are not sufficient to enable the cell to do what it needs to do. For instance, sometimes a cell needs to enhance or increase a concentration gradient because a high concentration of a particular molecule may be needed to create the conditions for a particular type of reaction to take place. I showed this in chapter 2 in the case of nerve cells, which have gates in their membranes that pump sodium and potassium ions up their concentration gradient so as to generate an electrochemical gradient—or an imbalance of concentrations of molecules and electrical charges—across the cell membranes. And for another instance, sometimes a cell needs to foment the making of a bond between molecules that might not easily find themselves in the same location or that might not readily interact: such molecules need to be brought forcibly close together so that the tension generated by their compelled proximity creates the conditions for them to bond. This is the case in cellular respiration, as elaborated in the pages ahead.

The creation of the conditions for these kinds of reactions—moving molecules up a concentration gradient and compelling their proximity—is counter to the tendency of energy-in-the-form-of-matter to persist in its least frenzied state. Consequently, a cell must use a process other than diffusion to move molecules up a concentration gradient or to have them come close enough together to interact with one another. This alternative process entails the production and use of high-energy molecules that attach to proteins and trigger energetically difficult work. Such high-energy molecules might prompt the opening and closing of the protein gates in cell membranes that transport ions and molecules in directions they would not otherwise go. Or they might effect the squeezing of reactant molecules together by proteins known as enzymes. In both cases, the work they do is called phosphorylation (pronounced foss-for-il-ation), and it is key to our exploration of the uses of oxygen in breathing.

Phosphorylation involves an extremely highly charged molecule called a phosphate group. A phosphate group is highly charged because, for various reasons, it has three more electrons than it does protons and those electrons are squished together in a way that makes the molecule energetically agitated.[1] In phosphorylation, one of these highly charged phosphate groups attaches to a transporter gate or an enzyme, both of which are composed of protein molecules. When a phosphate group attaches thus, its high energy level rearranges the electrical charge of the entire protein such that it changes shape. In other words, the excess of electrons that makes the phosphate group so energetic changes the resonance patterns and the bonding patterns of the atoms in the protein molecule.

This change in resonance patterns triggers a shift in the protein's shape that creates a movement akin to that of a gumball machine turning its coin-candy mechanism or akin to a fist clenching. This shape-changing movement in the protein thus phosphorylated facilitates the transportation of a molecule across the cell membrane or the squeezing-binding of two molecules together. So, in momentarily adding its frenzied energy to a protein, a highly charged phosphate group enables another biochemically important transition in energy.

Nelson and Cox (2008) claim that in an organism, and thus in a cell, a phosphate molecule is quite literally the key to transport mechanisms and reactions that are otherwise energetically "impossible" (495). But phosphate groups do not just float around to do their work willy-nilly. In order to be used, phosphate molecules are carried around cells by another molecule. Joined together, phosphate group and carrier, they are known as ATP, which we could glibly dub "all terrain power" for mnemonic purposes but only if we also remember that ATP is actually an acronym for its chemical components.[2]

What is important for our investigation here is that after an ATP deposits a phosphate group onto a protein, and after that phosphate group has done its transformational job to that protein and facilitated a reaction, it drops off the protein to which it is jazzily attached and floats away. In most of the cases we care about for now, we are left with a protein that has done a job, a carrier molecule able to carry a phosphate group, and a reusable but unattached and difficult-to-reattach phosphate group. Because the reattachment of the phosphate group onto the carrier molecule creates a molecule that is more highly charged than the two components separately—which is to say that it is an attachment that cannot be accomplished directly via diffusion—the big question is exactly how to attach a phosphate group back onto the carrier molecule in order to make it an ATP and thus enable it again to serve its crucial role as a catalyst and trigger. And it is a big question. Brooker et al. (2008) explain that the gates in nerve cells that pump sodium and potassium ions up their concentration gradient consume 70 percent of the nerve cells' ATP supply (101). In order for nerve cells to continue the reactions through which they enable organisms to perceive and respond to their world and their well-being, that supply must be constantly replenished. Garrett and Grisham (2010) observe that at any point in time, the average amount of ATP in an adult human body is around 50 grams—the rough size of a small lime; in order to supply the energy needed for the millions of reactions all the cells undertake in their work that keeps such an adult alive, each ATP molecule is recycled thirteen hundred times per day (68).

Clearly, the reattachment of a phosphate group to a carrier molecule to regenerate ATP is itself an "impossible" reaction. It is a process that demands an input of energy, since the two molecules do not easily or readily find one another and bind. But since it would make no sense to use an ATP to squash the component molecules together to create an ATP—it would give a net gain of zero ATP molecules—cells draw on an alternative source of energy and rely on an alternative many-stepped process. The alternative energy source takes the form of the food we eat, and the many-stepped process involves the oxygen in the air we breathe. The food that we consume and the oxygen in the air we breathe are used to regenerate or reactivate ATP molecules in the many-stepped process called metabolism or cellular respiration. In fact, the process of metabolism, or cellular respiration, is a successive and repeated transfer or rearrangement of energy—and it is in that rearrangement that the oxygen we breathe in does its work.

Metabolism as Appropriation of Energy

It is in the processes of metabolism and cellular respiration that it becomes particularly clear that we need to think of living bodies in terms of transitions in energy rather than added-together composed things. Recall that the bonds that hold atoms together as molecules are energy: electrons rippling through molecular orbitals, drawn toward the positive loci generated by protons and constrained in their pathways by the mutual repulsion exerted by their negative energy. This is the case for molecules in general; it is also (therefore) the case for the molecules that compose the organic matter that we ingest as food. In metabolism, the food that we eat is broken down into its constituent elements: carbon, hydrogen, and oxygen, among others. And the energy that formerly bound those elements together into, say, sugar or fat molecules is co-opted or appropriated to bind phosphate molecules to their carrier molecules so as to create ATP.

As I showed in chapter 2, a fat molecule—also known as a fatty acid or lipid molecule—looks a little bit like a long, multilegged bug. It is composed of a long chain of carbons, each with hydrogen atoms sprouting from its sides, with a "head" at one end of the chain composed of a cluster of carbon, oxygen, and sometimes nitrogen and phosphorous atoms. A carbohydrate molecule also consists of a chain of carbon atoms. However, in this case, each of the carbons has been "hydrated" with water (HOH). What this means is that rather than each carbon in the chain having two hydrogens (the figurative legs in the bug), each has instead a split water molecule: a single hydrogen (H) on one side and a hydroxyl group (OH) on the other—an arrangement that alternates along the line of carbons as the molecular orbitals allow. And whereas the chain of carbons and hydrogens in a lipid molecule is relatively stable and unreactive, the chain of hydrated carbons in carbohydrates is reactive: the polarity of the oxygen in the oxygen-hydrogen appendages make it possible for the charges in carbohydrates to be rearranged fairly easily so that the carbohydrates can react with other molecules or be broken down into smaller pieces.

Indeed, it is because carbohydrates are reactive that they are the predominant ingredients in metabolism. Fats, or fatty acids, can be broken down and their energy used by a cell, but that takes considerably more work. For the purposes of tracing the work of oxygen in cellular respiration, I will focus on carbohydrates.

The breaking-down-of-food reactions and the binding-phosphate-group reactions do not take place in just one or two steps. Rather, they are highly choreographed, multiphased reactions in which on the one hand the energy levels of the carbohydrate molecules are concertedly increased and pieces are snatched off until nothing is left and on the other hand electrons and hydrogen ions are moved around in a confined space until an electrochemical gradient is formed that is so great that a tiny protein mechanism is able to use diffusion to force the rebinding of ATP.

As I will explain further, the breaking down of a carbohydrate molecule into its bare elements takes place in two distinct sets of serialized activity—glycolysis and something called the Krebs cycle (also known as the citric acid cycle). After these, the attachment of phosphate groups to carrier molecules takes place through the squeezing action of a special protein. (I imagine it as something like the ordinary version of a superhero dashing into a phone booth and being squeezed into the superpower suit.) The oxygen we breathe is a critical factor in the squeezing process . . . So yes, it comes later.

Glycolysis. In the case of food, the mastication that takes place in the mouth and the mashing around in an acid bath that takes place in the stomach—these both break down carbohydrates into molecular components so that they can be absorbed into the blood. I do not want to work through the details of digestion here, although it is an amazing and fascinating process.[3] What I want to focus on is the fate of the glucose that is derived from the carbohydrates consumed and digested and that is absorbed into the blood, transported to cells, and metabolized.

Glucose is a carbohydrate composed of a string of six carbons, each of which has its requisite hydrogen (H) and hydroxyl (OH) appendages. In human bodies, glucose molecules are most often circular: one part of the molecule (the first carbon in the string, to be exact) bends around and attaches to another part (the fifth carbon), leaving the sixth carbon dangling out like a tail. It is because such a circular glucose molecule is too large and cumbersome to diffuse through the cell membrane that cell membranes have gated channels that are specifically designed to transport glucose across the cell membrane and into the cell.[4]

Glucose molecules travel into the cell via the power of diffusion, which is to say they travel down their concentration gradient—the glucose transporters simply make the travel faster. As noted in chapter 2, the first thing that happens to a glucose molecule on entering the cell is that it gets "tagged" with a phosphate group, which attaches to the tail-like sixth carbon. Lodish et al. (2008) note that this addition of a phosphate group to the glucose molecule is the first in a series of ten reaction steps through which the bonds and charges in the glucose molecule are rearranged so that it is able to be split into two identical three-carbon molecules each of which is a chemical called pyruvate (482). Why do they need to be identical? Efficiency: if they are both the same, then the cell needs only one set of mechanisms for the next cycle of breakdown rather than two (482). So, included in the ten-step series of reactions are those that open the tagged glucose circle, add a further phosphate group (to balance the molecule), and then shift the pattern of double bonds so that the molecule can be cut into two identical pieces and its resultant products cleaned up for the next process.

As a whole, this series of reactions is called glycolysis—the lysis or cutting of glucose. According to Lodish et al. (2008), the various enzymes and reactants that are used over the course of the process of glycolysis create the conditions for the production of four ATP molecules (482). Not bad for one glucose molecule . . . but certainly not as good as it gets. For the pyruvate molecules produced through glycolysis are transported into tiny energy factories inside cells that are called mitochondria. There, they are further broken down into their constituent bits and pieces, with the result that an additional twenty-eight ATP molecules are generated (485).[5] Let us look at that process in detail now.

The Krebs cycle. After a pyruvate molecule has been transported into a mitochondrion (singular of "mitochondria"), it is methodically taken apart. First, a carbon atom is ripped off by an enzyme. What remains is then picked up by an escort molecule and shepherded through a series of nine steps in which its carbon, hydrogen, and oxygen atoms are rearranged and peeled off until only the escort is left. The escort is then available to shepherd the remains of another pyruvate molecule through the very same process. Lodish et al. (2008) point out that this cyclic process is known as the citric acid cycle because the escort plus initial remains together constitute a citric acid molecule (487).

Now—and this is really important—the point of the nine-step breakdown of the remains of the pyruvate is not the empty but satisfied return of the escort to the beginning of the cycle. Rather, as Lodish et al. (2008) explain, the point of the ripping apart or dismantling is the deputation of two particular protein molecules to carry away the hydrogen atoms that had been components of the pyruvate molecule (487).[6] These protein molecules take the hydrogen atoms to another region of the mitochondrion. There, the hydrogen atoms are torn off the proteins, but in such a way that they become hydrogen ions (with a positive charge), while the electrons otherwise associated with those hydrogen atoms are passed into a series of way stations collectively known as the "electron transport chain" or the "respiratory chain" (493).

The electron transport chain is like a human bucket chain in which people pass along buckets of water to put out a fire—and in which each successive person issues more urgent and strenuous calls for the bucket to be passed: "We need water here!" In the case of the electron transport chain, each protein complex has at its center a large carbon molecule holding an atom that has a strong positive charge and that is therefore a strong electron attractor.[7] Each complex in the sequence is more attractive than the one before, so as each electron passes into one, it is drawn to the next by the force of its superior attraction.

At the end of this line is oxygen—the oxygen that diffused into the cell through breathing. Figuratively speaking, this oxygen has the loudest and most vehement voice. The electrical or energetic polarity of the oxygen draws the electrons out of the last protein complex in the electron transport train. In doing so, it creates the electrical room in those complexes necessary for the other electrons in the chain to be passed through. When the oxygen receives the electrons, it also snags a couple of hydrogen ions to create water—which then diffuses to be used elsewhere for other things. There it is. That is what oxygen does. (Does it feel like a letdown?)

But again, the point of the electron transport chain is not to get the electrons to the oxygen, although they do need to get there. Recall the ripped-off hydrogen ions? As the electrons pass from one complex to the next, their charged transit opens and closes portals that shunt the ripped-off hydrogen ions into a space between the two membranes that form the mitochondrion. As the electrons trip through the electron transport chain, the hydrogen ions are constantly shunted through the portals; they land in a milieu in which there are many other hydrogen ions. In other words, the shunting made possible by the passage of the electrons pushes the hydrogen ions through a membrane up their concentration gradient. And as the hydrogen ions gather in the space between the two layers of mitochondrial membrane, they create an electrochemical gradient across the membrane: lots of positively charged hydrogen ions on one side and very few on the other.

And here is the point of it all: the electrochemical gradient created by the hydrogen ions causes the positively charged hydrogen ions to stream down their concentration gradient (harnessing diffusion!) through a small protein passage in the membrane back into the inside of the mitochondrion. However, this passage is not a straight shot through a clear tunnel. Rather, the passageway is composed of the sections of a cog wheel—something like a playground's merry-go-round. As the hydrogen ions pass through the tunnel, they each serially push a section of the cog wheel, an action that turns a vertical rod at its center: round and round the rod turns as the hydrogen ions flood through. The rod is attached to the center of a protein that is a little like a revolving door—although in this case it is the rod that turns while the door, with its three sections, stays in place. As the rod turns, the three sections of the revolving door change shape: open, squeeze, open (Lodish et al. 2008: 503–9).

These rhythmically squeezing sections are pockets. The first "open" in each pocket is actually open-receive: the pocket receives into its fold a phosphate group and one of its soon-to-be carrier molecules. The squeeze phase squashes the two molecules together until they react to become ATP. The second open is open-release, in which the pocket opens to release the ATP molecule. So, as the hydrogen ions speed through the cog-passage, they rotate a protein complex whose changing shape squeezes and binds carrier molecules and phosphate groups into ATP molecules. In this particular process, the ripping apart of one glucose molecule will provide a gain of twenty-eight ATP molecules.

Just to review what the oxygen we breathe in is doing in this process: it is the last station in the electron transport chain, pulling out electrons from the last complex and pairing them with hydrogen ions to create water. It is only because oxygen is there, with its urgent attractiveness, that the passage of electrons progresses, the shunting of hydrogen ions across the membrane to create an electrochemical gradient continues, and the speeding of hydrogen ions down their concentration gradient that turns the cogs and wheels to create ATP. If oxygen does not arrive to draw and capture the electrons, clean away the hydrogen ions, and then leave as water, the entire process stutters, slows, and ultimately fails.[8]

Indeed, according to Lodish et al. (2008), cyanide—the notorious poison of captive spies who don't want to be compelled to spill the beans—is a tiny highly reactive molecule that, when it enters cells, has a particular affinity for the iron ion that is the electron attractor in the last complex of the electron transport chain before oxygen (498–99). When cyanide binds to that iron ion, it prevents the transposition of electrons to the waiting oxygen and thereby plugs up the entire electron transport chain such that the respiratory process stops—at that point, it doesn't matter how many breaths you take into your lungs, the cellular respiratory process ceases, your cells very quickly run out of the ATP they need to function, and you suffer a speedy death.

But to get back to breathing: the carbon and oxygen atoms that are disaggregated from the remains of the pyruvate molecules during the citric acid cycle pair up together to create carbon dioxide molecules. These gas molecules are small enough to diffuse through the various cell membranes into the blood, where they are eventually carried to the lungs. When the carbon dioxide–rich blood reaches the lungs, diffusion takes place again: the difference in concentration of carbon dioxide in the air and in the blood is such that the carbon dioxide in the blood diffuses down its concentration gradient into the air in the lung cavities. At the same time, the concentration of oxygen in the air and in the blood is such that oxygen diffuses down its concentration gradient from the air through membranes in the lungs into the blood. Once in the blood, oxygen attaches to four specially designed pockets on a molecule called hemoglobin that bears the oxygen throughout the body.[9] As the oxygen-rich blood circulates, it diffuses down its concentration gradient into the cells where it is needed. Its arrival and passage enables cellular respiration to continue—and permits life to go on.

Transition

Now that we have a sense of what oxygen does, it might be helpful to think about it a little further. The first consideration consists in a reminder of what it means to think about the body as energy-in-transition rather than substance-composed. What the example of oxygen shows is that while the body's cells do indeed compose and decompose substantial stuff, the point is not (necessarily) that the substantial stuff should have been composed or decomposed, which is to say that the material end product is not the point. To shift the focus of our attention here is not to deny or to ignore that there is substantial organic stuff that is composed or decomposed. It is, rather, to appreciate that the composing and decomposing activities constitute the (fleeting and sometimes temporarily stable) conditions for future cellular activity. In foregrounding the body as energy-in-transition, we highlight what the cellular activities of composing and decomposing make possible.

A second consideration concerns what I find interesting in the transience of oxygen. In focusing on its transience, I do not aim hereby to celebrate transition qua transition, to laud the transition in order to decenter substance, to prioritize the ephemeral because it undermines the solidity and immutability we attribute to substance. No, the transience of oxygen is interesting because the passage through which it manifests its transience has critical effects. The transit of oxygen facilitates the buildup and diffusion of energy in ways that enable cells to engage in otherwise energetically impossible biochemical reactions. The transience of oxygen reminds us that in a body conceived in terms of energy-in-transition, each transition always constitutes the unfolding of conditions for future transitions.

Moreover, the role that oxygen plays in effecting these transitions suggests that important molecules in living bodies sometimes gain their distinctive importance as a function of the living body's situation in a habitat rather than as a function of the particular activity that the molecule makes possible. When we examine the biochemical activities that together compose cellular respiration, we can see that each step lays down the conditions for the next. As a consequence, each step, each reaction, is integral to the overall process. To recognize that each step is critical to the integrity and movement of the entire process of cellular respiration would seem to entail that each of the successive steps or reactions are equal in the sense of being equally important: To have one step falter or fail is to interrupt the process as a whole. So, what would make some one molecule more important than another cannot be a function of its role or action in that process. In such a circumstance, how could we count the work of oxygen as more important or as the most critical?

Oxygen is not the "hero" of the cellular respiratory process. The particular reaction that is made possible by oxygen's passage is not appreciably bigger, trickier, or more critical than any or most of the other steps in the process of cellular respiration. In other words, we could not really say that oxygen is working harder or doing a bigger job.

The peculiar value of oxygen cannot be ascertained by considering its passage alone or even the specific reaction that its passage makes possible. We have to look more broadly. Oxygen is critical, crucial, terribly and vitally important because it is the step in cellular respiratory processes that is most profoundly dependent on the living body's engagement with its environment. And accordingly, it is the step that is the most vulnerable or the most susceptible to disruption.

To say this is not to deny that each of the other steps in a cell's respiratory process could be obstructed or disrupted. For example, a genetic problem might prevent the production of the various biomolecules necessary to run the process. If this were a problem acquired over the cell's lifetime, then the particular afflicted cell would likely die; if it were a problem with the organism's genes more generally, then likely the organism would not be alive in the first place. Alternatively, glucose, which is also dependent on the body's engagement with its environment, could be in short supply. However, this would not affect the body too much too quickly, because the body stores a version of glucose called glycogen in the liver to compensate in just such a contingency (to say nothing of the stores of fat molecules that can be ripped apart and, after some rearrangement, subjected to metabolic processes). The larger point is that with the interruption of other steps in cellular respiration, either the body has a backup resource, or a particular cell or couple of cells might die off while the organism as a whole could continue to persist.

The passage of oxygen in a cell is distinctive because it is not conditioned primarily by the composition of the cell and its biomolecular parts. Oxygen's activity is a system-wide, organismic contingency rather than a particular cellular contingency. It diffuses throughout the bodily system in order to have its particular effects in cells. Oxygen is important not for what it is or for its role in a process that has no constant object. Rather, oxygen is important by virtue of the fact that for oxygen to fail to pass through one cell is for it to have failed to pass everywhere. In other words, oxygen is an index of the profound dependence of an organism on the traffic and flow of very particular material conditions in its environment.

Oxygen calls on us to foreground theoretically not its role as an agent in a process but rather the fact that living organisms must take the world into themselves, ingest it, consume it, absorb it, dismantle it, appropriate it, and expel it. If living organisms are to survive, they must co-opt the energy-cum-matter of the world and incorporate it into the chemical and cellular activities through which the body-as-energy-in-transition is given form and persistence. In such a framework, the environment is not a context for creaturely living nor is it what Tuana (2008) might term a "viscous" and variably invasive field in which the activities of living proceed. Rather, the environment or habitat is the energetic condition, the condition and the constraint, for the transitions in energy that make creaturely life possible.

FIVE

Time

*In which we learn that the "it" of a
living body is a function of time*

Throughout this book, I have been trying to meet the challenge of elucidating some of the movements, processes, and logics, and some of the substantive details, attendant to thinking about creatures—including humans—as biocultural. Working from within a framework in which matter is conceived as an effect of energy that is constrained in its self-relation, the chapters have elaborated in various ways how a living body's biochemical activities depend on and effect transitions in energy, transitions that themselves are profoundly dependent on a living organism's immersion in and engagement with its habitat. This chapter addresses a puzzle that arises once we apprehend the extent to which a habitat provides the conditions of possibility for an organism's persistence in living. And, ironically enough, the puzzle concerns whether, in accepting the idea of biocultural creatures, we can imagine that a creature is at all or in any respect distinct from its habitat.

For, having elaborated how the selective permeability of cell membranes makes possible both the traffic of the environment into a body and the response of that body to that traffic, having argued that there is no feature, substance, or process in the body that is isolated or protected from that traffic and response, it has become unclear whether we can maintain conceptually that a living body is not merely an epiphenomenon of its environment or habitat. In trying to undo any residual sense that there is a biological material somewhere in a body that is insulated from the traffic of molecules in from and out to the environment, perhaps I have gone too far and lost a basis for distinguishing at all between a body and its environment. If a body composes and recomposes itself through transitions in energy effected and made possible by that traffic, if the body's activities are made possible by and are a response to that traffic, can we imagine the body having any role in directing, making possible, or delimiting such responses? As I showed in chapter 3, biological activities are directed because of the ways that precise prior steps constrain or delimit subsequent possibilities. If we are to stay true to such an insight, then we cannot suppose that just any kind of transformational response in an organism is possible in a given circumstance, as if living creatures were akin to fantastical shapeshifters. For instance, I cannot grow wings or extra limbs or extract oxygen from water or purr, even if I come to live in environments where such a recomposing response might be apt. In fact, Papoulias and Callard (2010) make just such a point, observing that an organism's responses to environmental stimuli do not follow an "aleatory bio-logic" but are instead "very precise, constrained courses of action" (41). If we follow Fausto-Sterling (2004) in her claim that "it is in our biological nature to generate physiological responses to our environment and experience" (31), and if we also hold onto this book's claim that what we think of as "biological nature" is biocultural, then how are we to figure or conceptualize what it is that conditions, constrains, and makes possible such very precise responses?

To frame the problem somewhat differently, perhaps in my effort to challenge the possibility of an überbiological substance, I have figuratively erased the body as an ontological entity, reducing it to not much more than a ripple in an active landscape. In posing the problem in this way, I am thinking of Judith Butler's early efforts to make visible the specifically political parameters through which what is considered "human" is rendered intelligible and materially real (1993; 2004). In an effort to expose how the discursive regime of heterosexuality circumscribes which forms of gender, sex, and sexuality can be counted and lived as properly human, Butler denaturalizes sex, arguing that what seems to be natural or given in and as the body is an effect of the performative citation of heterosexual norms of intelligibility. Seeking to foreclose the possibility of any reference to a pure, extradiscursive "nature" or "biology" that could ground a claim to the givenness or rectitude of heterosexuality, Butler contends that what we see and know as a sexed body is a discursive formation, a corporealization or materialization of historically specific norms of (human) subjectivity. As Butler (1993) herself notes, her argument about how embodiment is enframed within a heterosexual schema of intelligibility met with indignant protests by some among her critics on behalf of something in the body that is not simply and purely socially constructed (10). And of course, Butler concedes that there is a body that is "there" even if we cannot talk about it in ways that do not reinstall "nature" or "biology" as an extradiscursive reality (10; Frost 2014).

My work in this book is a similar effort to challenge the idea of a pure nature or biology. However, in reflecting on my attempt to undo the idea of an überbiological part of the body by elaborating the biocultural character of the processes that constitute living, I wonder whether I have merely substituted "environmentally constructed" for the notion of "socially constructed" and lost any notion of a body that is "there." If I have substituted "environmentally constructed," I have not really undone the binary distinction between body and environment but rather have just slid all the way over to the environment as the means for explanation. If the notion of a biocultural creature is to have any purchase, we will have to be thoroughgoing in our acceptance of the permeability of the cell membranes at the same time that we refuse any suggestion that a living organism is figuratively no more than a smear emerging from the material conditions of its habitat or environment. We have to account for the body's phenomenal and ontological "itness." The key problematic of this chapter, then, is to maintain the porosity of a living organism without either obliterating the organism altogether or, via a compensatory gesture, reinstituting some facet, dimension, or substance in the body that is not biocultural.

It is possible to capture in what such a distinctiveness might lie. In keeping with the idea of biocultural creatures elaborated in this book, we must turn away from a spatial framework that rests on a substantive understanding of the distinction between the inside and the outside of the body. We must cleave instead to the conceptual primacy of processes and activities, and we must think in temporal terms about the distinction those processes effect. So far in this book, I have imaginatively, if not self-consciously, suspended time in order to gain a richer and more textured sense of the permeability of and biochemical activity in an organismic body. However, the processes of energy transition, and of composing and decomposing, are continuous and constant: they are ongoing as living creatures live a life course. Accordingly, in this chapter we must suspend the suspension of time and insist on conceiving of each and every organism—and each and every generation of organisms—as coming into being, persisting and growing, and passing out of existence. If we endeavor always to consider organisms living lifetimes and reproducing in those lifetimes, then we will be able to appreciate that what makes any given organism distinctive vis-à-vis its habitat is a kind of temporal disjunction between the responsive activities it is prepared to undertake and the responsive activities provoked by its habitat.

As I have shown, when an organism develops and lives its life, it not only assimilates, appropriates, and engages with its habitat but also is modified through that assimilation, appropriation, and engagement. When an organism encounters, perceives, absorbs, ingests, and responds to its habitat, the encounter, perception, absorption, ingestion, and response results in changes to the kind and concentration of biomolecules that flow into and out of the organism's cells and that aggregate and react with one another. These fluctuating concentrations and varying kinds of biomolecules shape the production of the various proteins that facilitate and engage in the cellular activities that make continued living possible. We could say, then, that a developing and growing organism constitutes and reconstitutes itself, composes and recomposes itself, through its responses to features of its habitat. And, by virtue of the activity that the proteins facilitate, these responses delimit further responses at the same time that they make them possible. Put slightly differently, an organism carries or incorporates features of the habitat not simply in the form of influx of molecules but also in the form of its response to that influx—and its response to that response. Such activity-in-response is a constraint on as well as a condition of subsequent activity-in-response.

Crucially, some of the biomolecules whose production is elicited through an organism's response-to-habitat and whose activity transforms that organism can be passed on to offspring. Consequently, the bodily effects of an organism's habitat-engagement shape not only its own composing and decomposing but also the growth and development of its progeny in the next generation—and in the next. If we consider this transgenerational shaping from the perspective of that "next and next," we can see that an organism is the living trace—an accretion, a many-layered palimpsest—of many histories of creaturely engagements with habitats. An organism can be seen as a literal corporealization of a conjunction between its transgenerational carried history and the environment within which it currently lives.

This carried and corporealized history of response-to-habitat is extraordinarily significant for the purposes of this chapter, for that living history is the organismic material and activity through which the current environment has its effects. The proteins and other biomolecules that an organism's cells use to process and respond to chemical signals generated through the organism's encounter with the environment—those proteins and biomolecules are produced, in part, by the responses of previous generations to their environments. As a consequence, the effects that an environment can have on a porous organism are constrained by the responses of those previous generations to previous environments. Similarly, and from the obverse perspective, the responses an organism can muster in response to the provocations of its habitat are constrained by the ways that prior generations' responses shape its own processes of composing and decomposing. In short, the living history-of-response that constitutes an organism is the material and limiting condition of its responses to its habitat. So, while in some respects the porosity of living bodies could require that we say there is nothing in a body that has not been composed in response to environmental stimuli, the consideration of organisms in a temporal horizon precludes the possibility of positing an untrammeled arrow of causation composing and defining an organism from the noise of its contemporaneous environment. In fact, it is the long histories of response-and-adjustment-to-habitat that enable an organism to live in and meet the provocations of its extant habitat. The histories of habitat-induced responses through which an organism composes and decomposes itself mean that an organism is not wholly contemporaneous with its environment. Indeed, it is through conceiving of organisms as noncontemporaneous with their habitats that we can grasp conceptually both their porosity and their distinctness. It is through organisms' noncontemporaneity with their habitats that we can conceptualize what it means to say that they are biocultural creatures.

To get a sense of how to conceive of organisms as "histories-of-response-carried-forward" and thus as noncontemporaneous with their habitats, we first need to confront and challenge the recalcitrant belief that there is something in the DNA used in reproduction that is überbiological or insulated from the environment and the effects the environment has on an organism's living activities. Having dispelled the possibility of an überbiological anything, I will explore how the porosity of cell membranes enables the environment to shape how organisms develop from fertilized eggs through birth and growth to reproductive maturity. I will then trace some of the mechanisms through which that environmental shaping extends through and is realized in the development and growth of subsequent generations of organisms. In both cases, I will show that in living organisms, prior responses to habitat inputs are a limiting as well as enabling condition for further responses. The activities that distinguish the inside from the outside of a cell, or the inside from the outside of a living organism, have a history. It is this history, as a set of patterns of responsiveness and activity in and between cells, that sets a living organism apart from its habitat to an extent such that even as there is a constant traffic of chemicals and stimuli across the permeable cell membranes, the organism is not reducible to that environmental interchange.

Cell Membranes Are Porous: A Reprise

Like many scholars of culture and politics, contemporary scientists are trying to reconceptualize and rearticulate the relationship between organisms and their habitats. Questions about the extent and significance of the body's porosity are at the center of such efforts. As more scientists let go of the assumption that there is some facet of the body that is sequestered from environmental influence, they contribute to a growing acknowledgment and an emerging consensus that the development and transformation of organisms across generations is made possible and shaped by their porosity.

The fact that this is a growing acknowledgment rather than a long-held insight needs explanation. The short version of the explanation is that the thinking and research of scientists has tended to be shaped by the figuration of the relationship between inside and outside, between biology and environment, that this book challenges. In other words, if I can reference Keller (2010) again, even though "everyone knows" that genes and the environment interact, hypotheses, research designs, and intradisciplinary discussions have relied implicitly on a figuration of the body in which the work of genes takes place relatively secure from environmental interference. But, as I have shown in the course of this book, scholarship in epigenetics and developmental systems biology has begun to challenge and transform this figuration, elaborating how environmental factors traverse porous cell membranes to exercise profound compositional and regulatory effects on the processes through which bodies live. However, what Keller (2010) describes as the "mirage of a space" between genes and the environment has experienced something like a last stand in research on reproduction.

To be able to appreciate what is at stake in this "last stand" and in recent challenges to it, it is helpful to distinguish between somatic cells and germ cells. A somatic cell is the kind of cell I have been presuming in my discussion up until now: it is a cell that, with innumerable other cells, composes an organism. Generally, somatic cells are composed in specific and specialized ways such that they make the part of the body in which they are located the part of the body that it is. As I showed in chapter 3, a skin cell, a bone cell, a liver cell, a blood cell: the different shapes, textures, and activities of these cells derive from the kinds of proteins and other biomolecules the cells make from genes in response to chemical and hormonal inputs from the surrounding cells and, by extension, from the environment. In each kind of somatic cell, certain segments of the DNA that compose and bracket genes are appended with biomolecules—epigenetic markers—that make certain genes inaccessible to cellular transcriptional machinery and that thereby prevent their cellular "products" from being made. Other segments of DNA are appended with epigenetic markers that make them more accessible to the cell's transcriptional machinery, which ensures that the proteins and biomolecules that could be produced with other particular genes will indeed be produced. Both of these kinds of epigenetic attachments to DNA constitute a form of regulation of the kinds of biomolecules a cell produces, a regulation that ensures that the cell remains the kind of cell that its compositional location in the organism requires.

In contrast, a germ cell is a cell that is used to germinate or seed an entirely new organism. Rather than having its compositional possibilities circumscribed and directed by its location among similar specialized somatic cells, a germ cell has a fairly open, multifaceted potential trajectory. Once the DNA from a germ cell has merged with the DNA from another germ cell, the combination will divide and produce the somatic cells that will specialize and multiply to compose the various parts of a new organism. In this just-fertilized state, the germ cell combination is considered pluripotential, which means that the now-fertilized cell and its immediate progeny have enormous flexibility in what kinds of cells they will become—in the kinds and quantities of proteins and biomolecules each will produce.

The so-called last stand concerns whether the DNA in germ cells is überbiological or biocultural, whether it is sequestered and protected from environmental influence or whether is it shaped by and responsive to environmental stimuli. In a replication of the dynamic I discussed in chapter 3, in which the boundary demarcating the distinction between nature and culture was shifted figuratively inward to sequester the putatively überbiological DNA from the environment, germ cells have for a good while been considered effectively vaulted, protected and isolated from the environments or habitats in which the carrying organisms live. As Mayr (2002) articulates this classic position, "the genetic material" in germ cells "is constant ('hard'); it cannot be changed by the environment" (100). The idea here is that, as a consequence of being buffered from the environment, the DNA in germ cells is untrammeled or unsullied by habitat provocations and insults; the only thing that can disrupt it and cause a mutation is a mistake in copying as cells prepare to divide. This presumption about the "hardness" of germ cell DNA has manifested in accounts of reproduction and development as taking place unaffected by the vagaries of the environment. With this construal of germ cells, the environment is conceived mainly as the field within which an organism endeavors to survive and reproduce rather than as a contributing factor in its development, growth, and reproduction.

It is precisely such accounts of development, reproduction, and evolution that have begun to be challenged in contemporary biological research. As scientists are able to give increasingly detailed renditions of the molecular and chemical activities through which chemical and social environments (toxins, stress, and food deprivation, for example) affect the uses to which somatic cell DNA is put as creatures live and grow, they are also beginning to ascertain how those patterns of response are passed on through generations. In other words, they are coming to recognize that as in the case of somatic cells, germ cell membranes are permeable, germ cells are not buffered, and germ cell DNA is not sequestered.

As Via (2001) points out, this shift by scientists toward considering how environmental factors affect the processes through which germ cells participate in reproduction and development is quite recent. And it came through historically synchronous developments in work on evolution as well as in reproductive toxicology. I will look at the work on evolution in more detail ahead. But the conjuncture of that work with changing understandings of reproductive toxicology is both interesting and useful for understanding some of the stakes in the question about whether germ cell membranes are permeable or protected.

Cynthia Daniels (1997) argues that throughout the twentieth century, long-standing assumptions of white and upper-class women's vulnerability during pregnancy underwrote the exclusion of white women from laboring in occupations considered stressful or potentially hazardous—which of course had bearing on their pay equity and career advancement. The discovery in the early 1960s of the terrible effects of thalidomide on a developing fetus affirmed that vulnerability and through the 1970s and 1980s led to a research focus on women as primary vectors of environmentally induced fetal deformities. Emily Martin (1991) has done marvelous work showing how assumptions about femininity and masculinity have been transposed into scientific accounts of the activities of germ cells (i.e., the passive egg and the heroic sperm). Cynthia Daniels (1997) extends this work to show how such assumptions shaped research about reproductive toxicology in ways that reinforced this focus on the vulnerability of women's reproductive systems to environmental insult.

In Daniels's (1997) analysis of findings on reproductive toxicology, women's germ cells were perceived as particularly under threat from exposure to toxins because the reigning commonplace that the entire lifetime's supply of female germ cells (ova) are produced in a short period during fetal development led scientists to conclude that the ova would deteriorate in later reproductive life and thereby be especially vulnerable to damage from toxic exposure. In contrast, the story goes, male germ cells (sperm) are continuously produced over the lifetime of the organism, a productivity that generates a continuously renewed supply of shiny, new, robust, superenthusiastic, and invincible male germ cells for reproduction. Elaborating on these characterizations, Daniels (1997) explains that the presumption that only virile, healthy sperm could undertake the demanding activity of penetrating an egg had as its corollary "the presumption that abnormal sperm are functionally incapable of fertilizing an egg" (589). This sense that a damaged sperm would simply be an infertile one absolved men of any contribution to environmentally induced developmental distress in an embryo or fetus. In Daniels's analysis, this "all or nothing" theory of sperm health persisted through the 1980s amid political hysteria about "crack babies" and policies directed toward punishing and controlling the reproductive lives of (especially) lower-class women of color (589).

However, as characterizations of the process of fertilization changed in the 1990s to include not only the wiggling of a sperm to penetrate the membrane of an egg but also the inextricable gripping and absorption activities on the part of the ovum, it became evident that damaged male germ cells could very well participate in fertilization and carry into that process chemicals or defects that affect offspring development (Daniels 1997: 593; 2006). And perhaps undermining the "deterioration" thesis, White et al. (2012) draw on very recent findings to claim that both male and female germ cells are continuously produced through the lifetime of mammals (at least). Eventually, then, scientists' research findings showed that both male and female germ cell DNA can be affected by environmental toxins and thereby have an effect on the development of an embryo or fetus. But Daniels points out that, perhaps predictably, when female-mediated reproductive toxicology is discussed, women's "lifestyle" behaviors (drinking alcoholic beverages, using recreational drugs, smoking) are the focus of attention, whereas when male-mediated reproductive toxicology is discussed, men's exposure to environmental toxins through work and war are the focus of attention—even though both lifestyle behaviors and environmental exposures are likely to be similar for men and women (1997: 596).

As I will show in the next section, one of the primary insights to be gained from the acknowledgment that cell membranes in germ cells are porous is that the same kind of molecular traffic in and out of the cell that one sees in somatic cells also occurs in germ cells. According to Tazuke et al. (2002), there are "intimate interactions between germline and somatic support cells," interactions that "are required for normal germ cell behavior and differentiation" (2529). That is, environmental factors not only *can* but *must* pass in and out of germ cells. Germ cell membrane porosity is a condition of the many interconnected cascades of hormone, enzyme, and protein activity that initiate, regulate, and sustain germ cell cellular activity. One implication of germ cell membrane porosity is that when an organism absorbs and responds to its habitat, the absorbed substances and the biochemicals produced by the organism through those responses shape the production and activity of biomolecules in germ cells as well as in somatic cells. Environmental stimuli, chemicals, and toxins that are perceived, ingested, aspirated, or absorbed into somatic cells also traffic into and affect germ cells as they themselves develop.

To say that germ cells are porous in the way that somatic cells are porous could seem to be to say very little: if one accepts that somatic cells are porous it makes sense that germ cells might also be so. A porous cell would seem to be a porous cell by any other name. But in fact, the acknowledgment of the porosity of germ cells is transforming scientists' understanding of the mechanisms of development and growth. For if germ cell membranes are porous, then germ cell DNA are biocultural. And if germ cell membranes are porous and if germ cell DNA are biocultural, then not only do environmental stimuli contribute to and shape germ cell production and their self-sustaining activities. In addition, the effects elicited in germ cells by the environment in one generation are carried physically by those germ cells into the next. With the understanding of germ cell membranes as porous, the environment or habitat cannot be construed simply as the context within which a creature grows and reproduces or as just the energetic condition of a creature's living persistence. Rather, it must be also seen as a constituent of what creatures will and can become over time.

Biocultural Responses through a Lifetime

Earlier in the book, I indicated that somatic cells in living organisms respond biochemically to biomolecules flowing across cell membranes as well as to social stimuli faced or anticipated by an organism (Kültz et al. 2013; McEwen 2012; Slavich and Cole 2013). Such responses affect which kinds of proteins and other biomolecules are made so as to facilitate further biochemical activity and response; the responsive changes enable cells and the organism at large to perceive, engage with, and survive in a habitat. Since germ cells are also porous, it is not a stretch to imagine that biochemical responses induced by environmental factors would also have an effect on the genes that regulate the development of an organism from a single cell to a whole creature. And in fact, this is precisely what scientists are finding: the development of organisms from single cells into whole creatures is shaped by the environments within which the developing creatures and their parents live.

Let's examine the processes of development more closely.

Creatures who house their DNA in the nuclei of their cells reproduce themselves as organisms, that is, create offspring, via a process called meiosis (pronounced my-oh-sis). Meiosis is the mixing of two sets of parental DNA through the merging of two germ cells.[1] Generally speaking, after that mixing of parental DNA through meiosis, the fertilized cell then splits into two cells, which themselves split, and so forth, to create the many cells that will compose the new organism. This splitting is called mitosis. Mitosis is the process whereby the DNA in a cell is copied in its entirety and the resulting duplicates sequestered on opposite sides of the cell. The cell then splits down the middle into two identical cells, each of which has its own complement of identical DNA. Meiosis and mitosis sound very similar as terms, so a handy way to remember their distinction is that meiosis (my-OH-sis) creates a new Organism, whereas mitosis (my-TOE-sis) creates toes or other specialized appendages, organs, or tissues through the splitting and multiplication of cells.

Now, importantly, the development of an offspring from meiosis takes place not just because of the genes—as if the magic is in the mixing of the germ cell DNA. Rather, the capacity of fertilized germ cells to develop in such a way as to create a new organism depends crucially on the ways the environment in its many forms permeates the body. Gesturing to and reminding us of the critical importance of the environment for an organism's persistence, West-Eberhard (2005a) claims that the development of fertilized germ cells depends on the "presence of large numbers of highly specific environmental inputs: particular foods, vitamins, symbionts, parental behaviors, and specific regimes of temperature, humidity, oxygen, or light. Such environmental elements are as essential and as dependably present as are particular genes" (6547). Given the permeability of cell membranes, the importance of the environment or habitat for development would seem to be perfectly obvious. But sometimes things can be so obvious as to not be noticed or taken into account. Indeed, West-Eberhard observes that because scientists have implicitly regarded environmental inputs as the obvious background—they are "dependable environmental factors"—they are often conceived as *merely* background or as givens rather than as "powerful inducers and essential raw materials" (6547). In other words, they have tended to be dismissed because they are so obvious.

But challenging such a dismissal of the environment and its provocations, West-Eberhard (1998) notes that "observable properties of molecules, embryos, organs, cells, and behaviors . . . are related via specific mechanisms to genes" (8417). Which is to say that the traffic of water molecules, oxygen molecules, amino acid molecules, sugar molecules, ions, and hormones into and out of cells is critical to the processes through which genes are used in the course of development. Indeed, without that influx and efflux, the biochemical processes that make a germ cell divide and proliferate would stutter and still.

Some of these environmental factors are very obviously needed as raw materials and sustainers of germ cell life: sugar, oxygen, amino acids. But others are less obvious, and their effects on development have come to light only in the past decade or so as scientists have learned more about regulatory genes. Regulatory genes are genes that deal with other genes: they produce biomolecules and proteins that attach to DNA in a fertilized germ cell to facilitate the building of the proteins and biomolecules necessary for mitosis—for cell growth, cell division, and cell specialization or differentiation. I have noted that the traffic of chemicals and biomolecules into germ cells affects the biological activities central to their own growth and persistence. That very traffic also affects germ cell activity when they blend through fertilization and begin development. More specifically, the traffic of such factors into germ cells and their response to such factors change both how and when regulatory genes are deployed and how frequently and repeatedly the products of those genes can draw on the other genes.

In a fertilized germ cell, then, cell development—and the proliferation and differentiation of cells into an organism—depends not only on the configuration of the DNA that merged through fertilization but also on the other biomolecules and chemicals in the germ cells that derive from the maternal and paternal somatic cells. In a germ cell, as in a somatic cell, a multitude of proteins and RNA molecules are scattered throughout the cell body. Klug et al. (2009) explain that when parental DNA in germ cells merges through fertilization, the placement and distribution of those proteins and RNA molecules around the now-fertilized cell have a differential effect on which regulatory genes are deployed in the pluripotent cell-progeny of the germ cells as they first begin to divide. In turn, this differential deployment of regulatory genes shapes how the multiplying cells "specialize" in such a way as to determine the front-back axis of the developing creature (489). That is, the arrangement of proteins and RNA molecules in the egg when meiosis occurs directs which of the subsequent cells will become the "head" and which will coalesce to become the effective "tail." This initial arrangement of proteins and RNA molecules also affects the activity of regulatory genes in those subsequent cells, directing the production of specialized cells for the kinds of tissue structure and organ formation that are coincident with that head-tail axial determination—eyes here, toes there, and so on.

And crucially, while all this division, multiplication, specialization, and arrangement proceeds in a developing creature according to the various regulatory factors that were in the germ cells, hormones and steroids secreted by the gestating (parent) organism in response to its engagement with its habitat continue to traverse cell membranes or initiate chemical signals that affect how and when genes in its own cells are accessed and used to compose proteins and other biomolecules. Such increases or decreases in the influx of those hormones or steroids in the gestating (parent) organism can in turn change how the regulatory genes in the embryo or fetus are used. And, Klug et al. (2009) claim, sometimes such changes can have significant effects on the functioning or morphology of the developing creature (745–46). In other words, the processes through which a creature develops are responsive to its (gestating parent's response to the) habitat.

Such responsiveness is known as developmental plasticity—the term plasticity here designating a capacity for responsive change. According to West-Eberhard (1998), developmental plasticity is the ability of a developing organism to use its palette of genes differently depending on the different kinds of environmental stimuli to which the organism responds (8418). It does not involve a change in the genes themselves. Rather, developmental plasticity is the capacity to change how the panoply of genes is expressed.

And so, to give a more concrete specific example of how developmental plasticity works: West-Eberhard (2005a) explains that there is a kind of a stickleback fish that appears in two forms, a slender big-eyed form which lives and feeds in the variable depths of the body of lake water, and a stocky small-eyed form which dwells and feeds near the muck on the bottom (6546). They have similar genes in their cells. However, the kinds of foods available to the fish at each level of the water contain different chemicals and nutrients that, when ingested, affect the timing of the expression of regulatory genes. So, same genes but differently-timed or heterochronous expression of them. And because, as Klug et al. (2009) point out, proteins tend to work "in complex interconnected networks" (565), different timing in the use of regulatory genes can have ramifying effects, transforming how a whole range of proteins are built, pieced together, and function. In other words, heterochronous expression of regulatory genes during the process of development can lead to sets of quite drastic morphological changes—big or small eyes, stocky or slender physique.

This habitat-induced heterochrony in the expression of regulatory genes occurs not only in response to different chemicals ingested in a variety of foodstuffs. As Painter et al. (2005), Champagne (2011), Mychasiuk et al. (2011), and Mychasiuk et al. (2013) have found, it can also occur in response to chemicals produced by the body because of famine, social behavior, stress, and the presence of chemicals and toxins in the habitat that make their way into the cells of the body. The larger point, here, is that in a developing organism, environmentally derived or environmentally induced biomolecules permeate germ cell membranes. Those biomolecules not only make development possible. As West-Eberhard (1998) notes, they also organize and reorganize the expression of genes in ways that transform how that development will occur and how growth will proceed over a lifetime (8418). In sum, what a developing organism becomes is profoundly dependent on the effects and residua of the organism's absorption of and responsive engagement with its habitat.

Admittedly, at this point, it might seem that I am working against this chapter's aims. In elaborating how environmental stimuli and desiderata shape the activity of regulatory genes in development and growth, I have reaffirmed the effects of the environment on an organism's growth and persistence without providing for any resistance, resilience, or obduracy in the organism that inhabits that environment. In other words, in painting a picture of a developing organism as a product of a series of responses, I might seem merely to have confirmed the environmental reductionism

I set out to foreclose: if an organism develops through response, how can we conceptualize it as anything other than a response? However, once we appreciate how such environmentally induced developmental effects can be passed from one generation to the next, then we can appreciate how an organism is distinct vis-à-vis its habitat even as the porosity of its cell membranes renders it open and responsive to its habitat. The key to thinking about an organism-that-responds rather than an organism-as-response is the passage of time: because the modes and means of responding to habitat stimuli are passed on from one generation of organisms to the next, the range of possible activities an organism can use to respond to its habitat is in large part the range of activities that its progenitors developed from *their* responses.

Biocultural Responses through Intergenerational Time

As a creature develops, it produces the germ cells that will seed the next generation. Like a tripartite Russian doll that bears a doll that bears another, so a gestating organism holds a developing creature that itself grows germ cells for a next generation. This means that when a gestating creature responds to its environment, its response shapes offspring development and is also registered in the germ cells developing in that offspring. So, as Landecker and Panofsky (2013) note, when a pregnant creature is provoked by her environment, "three generations are impacted at once" (348). The developmental and growth trajectory directed by environmental provocation in one gestating generation becomes embedded in the next generation. It is in these intergenerational responses to environmental inputs that we can begin to see the "itness" of organisms as they compose and recompose themselves.

Developmental plasticity is increasingly identified as a key factor in the kinds of cross-generational transformations that contribute to evolutionary change over time. The effects of previous generations' developmental plasticity and habitat-responsive growth become the condition or basis for the current generation's development, growth, and reproduction. The biochemical molecules and the proteins that are the response-in-progress to the environment in one organism constitute the basis for the biochemical molecules and proteins that are elaborated and transformed through developmental plasticity in its offspring. In other words, an organism's developmental response to its habitat in one generation constitutes the condition and starting point for developmental plasticity in the next.

It is not yet clear to scientists exactly how changes in environmental factors might induce changes in the development of creatures cross-generationally. As mentioned earlier in this chapter, transgenerational change or evolution has conventionally been construed as due largely to random mutations in genes. However, evidence emerging in the past couple of decades challenges this convention, suggesting, as Champagne (2011) claims, that the environment—and organisms' responses to it—can induce transgenerational changes that result in the development of new species (4). The permeability of cell membranes in both somatic cells and germ cells enables the influx and efflux of chemicals vis-à-vis the environment such that a biological organism composes itself in accordance with cross-generational embeddedness within and responses to habitat. If we look at some of the ways such transgenerational transformation occurs, then we can apprehend how a living organism carries response-traces of the many habitats its progenitors have engaged.

Research concerning the relationship between developmental plasticity, transgenerational change, and evolution is at the center of one of the key and rapidly changing debates in evolutionary theory. Broadly speaking, there are two camps in this debate. A quick sketch follows.

On one side is the long-standing argument that the physical isolation of a group so that they breed only among themselves enables a distinctive set of features to evolve in that group such that a new species is formed (Coyne 1994; Mayr 1999). The key formulation here is that reproductive isolation precedes differentiation.

Proponents of this view maintain a substantive distinction between inside and outside the body such that species are understood to evolve when a random genetic mutation (inside) enables a creature to survive and thrive in hostile or new environments (outside) in such a way that it can breed and pass on its genetic information to its progeny. Here, the environment is the field for the reproduction and survival of creatures whose slowly and randomly mutating DNA produces functionally useful proteins and biomolecules that make those creatures distinctive as a population over time.

On the other side of the debate is a newly rising consensus that draws on insights about the porosity of somatic cells and germ cells to claim that the environment induces change as creatures develop. Such environmentally provoked differences are passed on through generations via epigenetic markers. Then, through mechanisms akin to reproductive isolation, those differences eventually lead to the formation of new species (West-Eberhard 1998: 8419). The key formulation here is that differentiation precedes reproductive isolation.

In this view, the changing creature and its habitat are not related as object to field. Rather, creaturely differences develop in response to environmental stimuli. Champagne (2011) explains that developmental plasticity enables "cues regarding the quality of the environment to shape offspring development in such a way that they are better prepared to survive and reproduce in that environment" (10). In other words, developmental plasticity is a preemptive preparation for the habitat such that offspring "will not have to experience the threat themselves" in order to be biologically suited to deal with it (10).

While the first view has been dominant for many decades, with the second increasingly finding evidence to support its counterclaim, recent studies suggest that the modes of transformation staked out are likely not mutually exclusive alternatives but, as Jiggins (2006) proposes, "probably lie somewhere in between these two extremes" (334). What is at issue in this debate is not simply the question of whether the environment is a contributing factor in evolutionary change. More critical is the question of what is to count as evolution. If we examine the arguments in more detail, we can gain a sense of how the environmentally responsive activities of every organism are the accumulated and variously transformed histories-of-responses-to-habitats of earlier generations.

When genes mutate in a population that is isolated in such a way that reproduction occurs largely between within-group partners, then over time the mutation leads to the development of a new species. The details of this process center on the effects of random changes in DNA that result from miscopying or some kind of damage in the course of DNA replication. The idea here is that when the DNA in a germ cell is copied so as to allow the cell to divide, pieces of the DNA may be miscopied such that a nucleotide is missing, added, or simply substituted. If such a mistake occurs in a germ cell as it initially develops, the mistaken DNA will be copied when the germ cell begins to divide. Subsequently, the mistake will be copied into all the cells that derive from that original cell whose replication resulted in a mistake. As I showed in chapter 3, a miscopying of DNA can mean that a different amino acid is included in the resultant polypeptide, a difference that may affect the folding, the topography, and the function of the resultant protein. Or it may mean that the DNA regulating the expression of genes is altered. Whether the alteration precipitates a small or a large change in the development and growth of the organism depends on the particular genes, the particular kind of alteration, and the network of proteins affected by the change specified by the mutation.

If a random mutation results in a gene that is expressed more or less frequently than the premutation version, or if it results in the production of a slightly different version of a protein that maintains its function but alters the frequency, likelihood, or effectiveness of its activity, then the two versions of the gene are called "alleles." Alleles are "alternative forms of a single gene" (Klug et al. 2009: 45). To explain via reference to my fictive sample genetic sequence MMMMTHEMMMMFATMMMMCATMMMMSAT from chapter 3: in two alleles of a gene, the DNA that codes for particular kinds of proteins (THE FAT CAT SAT) could be the same. However, in one creature there might be four Ms preceding the sequence, whereas in another there might be nine. And the different number of Ms in each case might promote a different repetition in the use of the genes. For instance, the creature with the four Ms might produce THE FAT CAT SAT proteins once every time the biomolecules indicating need arose, whereas the creature with the nine Ms might make THE FAT CAT SAT proteins three times every time the biomolecules indicating need arose. Because the genes are roughly the same but the regulatory genes modifying the use of them are different, the two working versions or sequences are known as two alleles of the same gene. And because the proteins or biomolecules produced make specific biochemical activity more or less likely or efficient, allelic variations can make for slight differences among otherwise similar creatures.[2]

However, if a gene mutation leads to a protein that has a different amino acid sequence, that consequently folds differently, and that therefore performs a different function, then its effects on the creature—if it doesn't kill it—can be quite profound. Because proteins work in conjunction or series with many others, a protein produced through a gene mutation can shift the activity of the panoply of proteins and biomolecules with which that protein used to work. Similarly, if a gene mutation produces a differently composed RNA molecule—one of the biomolecules that work with proteins to facilitate the various activities of the cell and that compose the machinery that transcribes and translates DNA—then it can lead to a range of deactivations and reactivations of other genes, changes in timing of development, and wholesale transformation of cellular structure or function. Depending on what the cellular activities were that the older version of the gene used to enable to occur—and on what the cellular activities are that the newer version of the gene enables to occur—a mutation can lead to the formation of a differently characterized organ or physical feature (size, texture, strength, reactivity) or to the formation of one that is differently placed or oriented (top, bottom, underneath, to the side, toward the back, etc.) or has a different function. If such a mutation is copied and recopied as the organism and its progeny multiply over generations, and if the distribution and the effects of that mutation among a population are not diluted through reproduction with "out-group" members, then the mutation and the physiological features it provokes can become a repeatedly appearing feature of that population. And eventually, voilà: a new species.

Complicating such a tidy story, however, Bush (1969; 1994), Via (2001), and Wolinsky (2010) point to research findings that increasingly call into question the way this theory both limits the causes of change to random mutation in genes and circumscribes the effects of the environment on organismic development and growth. What is at issue is not whether geographic isolation can lead to the development of population-specific transformations—this is quite well established. What is at issue is the sequencing of events. Nosil (2008) observes that since the porosity of germ cell membranes can lead environmental stimuli to induce heritable developmental transformations in organisms, differentiation can and often does precede reproductive isolation. Similarly, West-Eberhard (2005a) points out that "environmental factors constitute powerful inducers and essential raw materials whose geographically variable states can induce developmental novelties" that appear in subsequent generations (6547).

As I have shown, the capacity for developmental plasticity means that environmental stimuli can change how an organism's regulatory genes provide access to the genes needed for particular features of an organism to grow. As creatures absorb and respond to the environments they inhabit, some genes that might heretofore have been transcribed first and frequently in the course of development might be switched off—or their involvement delayed—while other genes that might have been inaccessible for generations are marked for transcription. In other words, as West-Eberhard (2005a) explains, an organism's response to the environment can produce developmental cues that effect a kind of "mixing and matching" among different genes to make different versions of the proteins with different functions in different cells in the course of development and differentiation (6546). Extending this insight, Guerrero-Bosagna and Skinner (2009) and Skinner et al. (2013) contend that because the environment or habitat works its effects on organisms in various phases of development—from the generation of germ cells through fertilization and prenatal and postnatal growth—environmentally derived hormones, chemicals, and behaviors can shape how and in what ways a creature and its offspring might grow and transform.

Scientists are still identifying the mechanisms that allow developmental plasticity to result in the formation of a new species. For example, West-Eberhard (2005b) suggests that if environmental change can initiate genetic mixing and matching during development, then stability in that changed environment would entail the same kinds of genetic recombination from one generation to another: as each generation is exposed to that environment, each will develop in similar ways.

In addition to the effects of consistent environmental stimuli on multiple generations, a single event experienced by an organism in the environment could provoke the production of epigenetic markers that permeate the germ cells in such a way that they are produced and reproduced in the organism's progeny. Painter et al. (2008) argue that such has been the case for descendants of people who were born to women who were pregnant and malnourished during the Dutch famine in World War II. Guerrero-Bosagna et al. (2005) argue that genes appended with biomolecules that affect their transcription and translation into proteins carry or conserve the pattern of those epigenetic appendages "from one generation to the next" (344). This carrying forward or conservation occurs in part because while in somatic cells such epigenetic biomolecules generally are found in the nucleus, in fertilized germ cells as well as in early embryos, those epigenetic biomolecules are found throughout the cell body (343). What this means is that environmentally induced changes in development in one organism can modify how and when genes are transcribed in offspring. So, as new creatures develop, they use not only information from genes themselves but also information and chemicals deriving from the prior generation's engagement with the environment. In their gloss on this process, Guerrero-Bosagna et al. emphasize the ways responses to particular features of the environment are registered and passed on through germ cells, explaining that reproduction entails "conservation in the progeny not only of the structure required to carry out the self-conserved organization represented by the organisms but also the preservation of the structural characteristics of the environment that allow such organization to take place" (345).

Reconsidering Evolution

Notably, the possibilities outlined here concern changes in the ways genes are used—changes in how development occurs—but not changes in the structure or sequencing of the DNA itself. In the dominant view now under challenge, the development of a new species, or what is called speciation, requires changes in the DNA itself. According to Guerrero-Bosagna et al. (2005), West-Eberhard (2005a), and Nosil (2008), this consensus is being reconsidered because it appears that the kinds of developmental plasticity explained here could lead to changes in DNA. In other words, new kinds of genetically distinct creatures can emerge through intergenerational responses to habitat.

The possibility that developmental plasticity could lead to changes in DNA turns on the presence and placement of particular kinds of epigenetic molecules that modify the process of DNA transcription. The particular epigenetic molecules at issue here are methyl groups—a molecule composed of a carbon atom with three hydrogens attached. As is the case with many epigenetic molecules, the presence of methyl groups in somatic cells and germ cells is dependent on the response of an organism to environmental stimuli or stressors. Methyl groups in particular prevent transcription machinery from latching onto DNA; this interrupts the use of that DNA in the production of the particular proteins and biomolecules that make a particular kind of cellular activity possible. Methyl groups are found in germ cells as well as somatic cells, which means that the effects they have on DNA transcription will be passed on from one generation to the next. Guerrero-Bosagna et al. (2005) speculate that because methyl groups have a very specific chemical composition, their persistent presence across generations could precipitate changes in the DNA molecules that constitute genes.

Recall that there are four types of DNA—cytosine (C), guanine (G), thymine (T), and adenosine (A), each of which has a pairing preference such that C-G and T-A are reciprocal pairs in the double helix DNA structure found in cells. Methyl groups tend to attach to cytosine portions of the DNA. Guerrero-Bosagna et al. (2005) point out that when a methyl group is attached to a cytosine, the combined set of methyl-group-and-cytosine is chemically halfway toward converting to a thymine: only a small chemical reaction is needed actually to make that set a thymine (344). And, given the C-G and T-A pairings that are so central to the precise copying of DNA when cells reproduce, to have a thymine substitute in a DNA sequence for a cytosine is to provide for the possibility of numerous different amino acids in a sequence aimed at producing a protein (and a number of different outcomes in a ribonucleotide sequence intended for a functioning ribosome of some sort). Guerrero-Bosagna et al. speculate that if there are persistent patterns of cytosine methylation from one generation to the next, then the likelihood of the reaction occurring that would transform the DNA is relatively high (344). What this means is that an organism's sensitivity and responsiveness to its environment, as manifested through developmental plasticity and epigenetic modification, can lead to changes in the structure or sequence of DNA over time.

And of course, in order for a transformation to become a feature of a population rather than just one or a number of individual creatures, it has to be reproduced and distributed through a population over time. Importantly, if differentiation can precede reproductive isolation, then the contours and features of reproductive isolation are not limited to geographical separation. As developmental changes accrete and expand over time, as their effects combine in various ways, then creatures who might in a previous generation have been similar can become so dissimilar that they can no longer reproduce together. West-Eberhard (2005a) suggests that such an incapacity to reproduce might derive from an acquired incompatibility of the parental germ cells and DNA—in the sense that the germ cells will not provide the biochemical precursors, conditions, markers, or developmental signals that enable a fertilized cell to divide or to flourish and develop into a new creature (6548). Alternatively, physical size or other morphological differences among creatures can make reproduction technically impossible (whoops!), or differences in ages and duration of reproductive maturity can mean that they are no longer synchronized in such a way that reproduction is possible (6548). Since only those creatures who were reproductively compatible would be able to reproduce, the developmentally induced difference could become a population-wide divergence.

One of the implications of the research that supports the thesis that differentiation can precede reproductive isolation is that creaturely transformations and the eventual genetic changes that lead to speciation do not occur just randomly. Rather, because they occur in response to particular features of or substances in the environment, they can occur in patterned and particular ways as organisms engage with and respond to their habitats. In other words, creaturely transformation or species evolution is closely indexed to features of and events in the material and social habitat. If the environmental stimulus that precipitates developmental changes is shared or experienced by a group of creatures, then the changes could be patterned and fairly widespread rather than having to originate from a single creature with a random genetic mutation. This in turn provides for the possibility that a population-sized differentiation could take place over several rather than many hundreds of generations.

Another consequence of this research is that it enables scientists to account for the appearance of similar genes across many different species. West-Eberhard (2005a) explains that because developmental plasticity enables "the same genes [to be] used over and over in different contexts and combinations," a population of creatures could diverge quite significantly, become reproductively incompatible, and speciate while at the same time maintaining some genetic similarity (6548). Indeed, Barlowe and Bartolomei (2007) propose that genetic similarity among divergent species is in part an effect of an epigenetic mechanism whereby genes that are interrelated and whose protein products work especially well together are not only chemically bound together during germ cell fusion—preventing mix and match—but also are protected from epigenetic reprogramming so that, as a set, they are available for use in the development of further generations. In other words, epigenetic factors are responsible for preserving or conserving particular DNA sequences as well as for modifying or transforming others in ways that lead to differentiation and species divergence.

This kind of preservation of DNA sequences is actually a really interesting phenomenon. As I have shown, each cell has many kinds of proteins that do many kinds of tasks. Some give shape and structure to cells, some make it possible for biochemical molecules to traffic in and out of the cell, others trigger reactions or make reactions possible. And it turns out that while genetic and epigenetic changes over time have transformed which proteins are produced, when, and with what frequency, and while those changes affect various of the features, functions, and appearances peculiar to particular creatures, other proteins have not changed much at all. To tweak a common saying: when some things change, others stay the same. So, the proteins needed, say, to create a water pore or a gated sodium ion channel in a cell membrane have changed very little even as developmental plasticity and random mutation together have enabled other transformations to occur over millions and millions of years. If a cell is a cell is a cell—roughly speaking—then there are some proteins whose composition and shape enable a particular activity or process to occur in a cell no matter what the creature. The genes that code for such proteins escape epigenetic reprogramming, avoid mutation, and are possibly protected in such a way that they can be used by manifold generations through evolutionary history. Thus, Klug et al. (2009) point out, in cells from yeasts, fruit flies, flatworms, and zebra fish and on to mice, rats, and humans, there are functional proteins and related gene sequences that are roughly similar (486). When the functions of proteins are roughly similar among creatures, and when the gene sequences that enable the production of those proteins are roughly similar, we say that the proteins, their functions, and the genes that are the template for them have been conserved.

Incidentally, it is because there are genes and proteins that have been conserved over the *longue durée* of evolution that yeast, fruit flies, and other little creatures are considered "model organisms" for scientific study. Such little creatures reproduce very quickly and prolifically, which means that they show the effects of epigenetic factors and genetic changes quite speedily. This enables scientists to gain a good sense of how particular genes, epigenetic markers, and proteins work—in the small creatures and in larger creatures, including humans.

West-Eberhard (2005a) suggests that a further implication of the research indicating that differentiation can precede reproductive isolation is that "the evolution of a divergent novelty does not require gene-pool divergence, only developmental-pathway and gene-expression divergence" (6548). That is, since epigenetic changes can be passed on for generations in a stable environment, we can expect to find differences among creatures that are so significant that the creatures appear to be different species when "they are in fact complex alternative forms that represent gene-expression, not genetic, differences between individuals" (6548). Indeed, Lordkipanidze et al. (2013) claim that such is the case in recent findings that what had been considered genetically distinct prehuman homo species were actually genetically the same: rather than being distinct species, they were different morphotypes . . . which is to say that they just looked different. Guerrero-Bosagna et al. (2005) draw another conclusion from the findings that developmental plasticity can lead to changes in organisms that are consistent and persistent over generations, suggesting that "the definition of 'evolutionary change' . . . becomes blurred" (348). In other words, they claim, if we find "that persistence [of changes] through generations could be achieved in alternative ways to genomic mutation," then our definition of evolution itself may have to change (348).[3]

Noncontemporaneity

What is important for our puzzle in this discussion is the ways that a living organism's processes of composing and decomposing, its intra- and intercellular activities that effect transitions in energy, and its modes of responding to environmental stimuli are temporally layered. In any one creature, long-term cross-generational priming-to-respond—in the form of DNA, epigenetic biomolecules, and ecological factors—combine with near term developmental plasticity and current habitat to constitute that creature's particular modes of responding to habitat. It is because of the ways that past habitats are traced in those cross-generational templates for how to respond, build, or decompose that I want to say that a living organism is noncontemporaneous with its current habitat.

When a creature responds to, let's say, Habitat A by producing biochemical or epigenetic markers that affect gene expression, those markers are also appended to germ cells. The presence and work of those markers in germ cells means that the second generation is shaped by features of Habitat A—even if the second-generation creature does not live in that habitat. The proteins and cellular activities composing and decomposing the second-generation creature are made possible and constrained by the effects of Habitat A on the somatic and germ cells of the prior generation. The biomolecular, cellular, organismic responses of the new generation to its own habitat—let's say, Habitat B—are conditioned and circumscribed by the prior generation's responses to Habitat A. The ways that the new generation is able to develop and grow in response to Habitat B bear the trace of the prior generation's response to Habitat A. Indeed, the new generation can live responsively precisely because of the way that the prior generation's modes of responding are replicated as a template in their cells. With so much of its composing and decomposing oriented to responding to Habitat A, the new generation living in Habitat B is in a response-lag.

If we extend such a carrying-forward of a prior generation's response through multiple generations, we can see that, as a product of the many generations preceding it, an organism's habitat-responsive development and growth is noncontemporaneous with its current habitat. Those histories, as they manifest in cellular structures, protein activities, and organismic processes, set the organism apart from its habitat in the sense of providing the possibilities for and the constraints on the biological responses that the habitat can evoke or induce. In other words, the fact that the effects of porosity endure over time, through creaturely life cycles, and across generations means that an organism is composed not merely through its encounters and engagement with its contemporaneous habitats but instead through the conjunction of the accumulation of prior responses-to-habitats and its responses to its contemporaneous habitat. It is this corporeal history, this noncontemporaneity, that gives a living organism its "itness" even as it composes and recomposes itself continuously in response to and through engagement with its habitat.

What is interesting in this understanding of organisms as noncontemporaneous with their habitats is that it enables us to imagine the responsiveness of an organism to environmental stimuli while avoiding the stodgy Charybdis of inviting a behaviorism redux and the unsatisfying Scylla of imagining an indeterminate infinity of emerging possibilities. In the view proposed here, the activities and processes through which an organism responds to its absorption, ingestion, and perception of its habitat are made possible, conditioned, and constrained both by how past generations have responded and by what the current habitat provides in the way of provocations. Because of this temporality, we can talk about a specific organism *responding* to an environmental stimulus—as if not equivalent to it—without reinstalling a sense of überbiological, substantive separateness from it.

Conclusion

To my mind, theory is a collective if not a collaborative project, an internally motley and sometimes viciously contestatory endeavor. What I offer to that endeavor here is not a polished and sealed theory of the human but rather a set of concepts, figures, and reminders, a vocabulary of movement, a terrain of possible resources. In closing, I would like to sketch some further thoughts that anticipate how the idea that humans are biocultural creatures might be elaborated.

Humans as Biocultural Creatures . . .
To consider humans as biocultural creatures is to have a basis for thinking about humans as political subjects without recapitulating the forms of human exceptionalism that have relied on a disavowal of materiality, embodiment, animality, or dependence. I have delved down into the quanta, the atoms, the molecules, and the cells not because I think it is important for us to think and articulate our political ideas in idioms pertinent to those scales of existence. Rather, I have dug down through them because in doing so I have been able to do two things. First, I have been able to dislodge the notion that there is any aspect of a living organism that is

not cultured, that persists on its own accord rather than through its interaction, interrelation, and transformation with other substances, creatures, and organisms. And second, in playing around with and playing out the idea that living matter persists in its living by virtue of constraints on and transformations in energy, I have been able to undermine the substantive binary distinction between body and environment while also maintaining a sense of the distinctness of the living organisms cultured by their environments. This distinctiveness is important, I think, because it lends itself to an affirmation—however attenuated it might be—that humans can engage in deliberate collective action.

The idea that humans are biocultural creatures provides a basis for developing richer and more detailed accounts of how social and material worlds are constitutive of the way we live and experience our lives. To put the point more bluntly, it gives us the conceptual language and figures to articulate how ideas and culture shape living matter. The conceptualization of matter as an effect of energy under constraint enables us to trace how the processes of sensory and imaginative perception generate energetic chemical reactions within and between cells that transform the functioning of the cells, the responses of the endocrine systems, and the patterns of the broader anticipatory composing and decomposing of the biocultural body. It enables us to apprehend how a biocultural subject's conscious or nonconscious perception of something in the self or beyond the self could produce a response that becomes, for some period of time, a part of the biocultural self, a part that constrains, directs, delimits, and also makes possible other kinds of perceptions and responses.

Developing this thought further, the concept of biocultural creatures enables us to think of humans as perceptually responsive at cellular, organic, regional, and organismic levels, in ways that include neurological stimulation by social and material phenomena (i.e., sense perception and imagination) and in ways that also include more diffuse hormonal or biochemical responses to the ingestion and absorption of the material dimensions of habitats. With this broad understanding of responsive perception in biocultural creatures, we can trace in richer substantive and theoretical detail the incorporation of self-understanding and self-identity, the embodiment or materialization of individual reactions to the pull and push of various and possibly crosscutting norms and expectations—local and parochial as well as broadly shared. The idea of biocultural creatures provides a gestalt through which we can see that what theorists have termed the materialization or incorporation of norms is akin to what Boyce et al. (2012), Lock (2013), and Niewöhner (2011) describe as the "biological embedding" of social perceptions that derive from the striations, tensions, and inequalities in social and political life. In other words, with the idea of biocultural creatures, we gain a means to explain and to figure how the social, symbolic, and material provocations of collective life "are transformed into the biology of the body" (Landecker 2011: 178).

To entertain the possibility that the processes of materialization and incorporation that have been central to a range of theoretical analyses are of a kind with the processes of biological embedding is not to cede ground to a more nicely dressed biological reductionism or a poorly refurbished form of phrenology. For an experience or a norm to become incorporated, to become materialized in and as the self, or to become biologically embedded is not for it to become a fixed feature of the biocultural creature in question. Biocultural creatures are constantly composing, decomposing, and recomposing in response to their engagement with their habitats. The recognition of such de/re/composing activities requires that we foreground the temporalities of a creature's responsiveness to habitat, the sometimes fluid and sometimes disjunctive processes of responsive development, dissipation, intensification, and transformation. Experiences that become embodied or embedded do not become permanent features of the biocultural self, obdurately stuck and thereby determinative of all future forms and modes of responsiveness. The noncontemporaneity of biocultural creatures' responses to their varying habitats means that persistence in a feature or a form need not collapse conceptually into inertia, that repetition need not slide conceptually into determination. Within this framework, then, the materialization or embedding of experiences might be construed as a readiness to respond. Indeed, if we figure biocultural creatures as "seeded with traces of [their] future," as Maurizio Meloni and Giuseppe Testa suggest (2014: 445), if we figure them as anticipatory forms, then we can propose that the materialization or biological embedding of norms and social experiences is preparatory; it prepares the conditions for future responses that, in being experienced, will recompose the biocultural body anew. To account for the ways that a biocultural creature's embodiment of a response to its habitat is anticipatory is to be able to foreground the formative, compositional intermingling of creature and habitat while also not losing sight of that creature—that self—as a distinct being.

Importantly, however, we must acknowledge that the different spatial and temporal scales at which biocultural creatures respond to their habitats mean that what recomposes constantly at one temporal frame and one scale of reference may appear to be fairly stable at another. Even as we recognize that at micro-scales of time and space, biocultural creatures constantly undergo a frenzy of biochemical activity, transitions in energy, movements and shifts of diffusing molecules, and all manner of traffic across each and every cell membrane in the course of engaging and responding to habitats, we must remember that at the gross spatial and temporal scale of the organism, the creature, the human individual, the micro-adjustments and transformations might not be made manifest until they accumulate sufficiently so as to affect the larger scale function and behavior.

Another way to make this point is to note that even as the biocultural body constantly recomposes, that recomposition is not an unconstrained, free-form kind of transformation (Gunnarsson 2013; Papoulias and Callard 2010). The conditions of possibility for biocultural creatures to be responsive to their habitats are also, at the same time, a constraint or a delimitation. We can also say it the other way: the constraints and delimitations on how biocultural creatures might respond to their habitats are also the conditions of possibility for their responsiveness. In fact, this insight about the productivity of constraint has run through the argument of the entire book. It is because there are constraints on what carbon can do that it can be a generative element in living organisms. It is because there are constraints on how various molecules can interact that cell membranes are porous and that molecules diffuse. It is because those constraints effect the selective permeability of cell membranes that there is a constant traffic of stuff into and out of cells and bodies, a traffic that provides the conditions for the biochemical activities that keep living creatures alive. And it is because those biochemical activities provide the means and modes for an organism's responses and future responses to that influx and efflux that creatures persist, transform, and evolve into different kinds.[1]

Biocultural creatures are anticipatory forms whose creative responses to the provocations of habitat draw on a rich and deep histories-of-responses. The different modes of responding over and through time not only make an organism a distinctive "it" vis-à-vis its habitat. Those accumulated and variously changing modes of responding also differentiate organisms from one another. In being prepared by both their short- and long-term histories to respond to some features of their habitats, they are also—thereby—not prepared for others. Some biocultural creatures develop to respond in some ways to habitat stimuli, and others to respond in other ways. Some biocultural creatures, like humans, will have distinctively interrelated modes of responding, and other biocultural creatures will have some other distinctively interrelated modes of responding. In rehearsing what may seem very obvious here, I am suggesting that the idea of biocultural creatures enables us to talk about human biocultural creatures as adaptively and reproductively distinct from other biocultural creatures without at the same time denying their biocultural-ness—their materiality and corporeality, their frenzied vitality and animality, their finitude, vulnerability, and profound dependence on the habitats and the other biocultural creatures that culture them (Haraway 2008; Hird 2009b).

To sum up, then, the idea that humans are biocultural creatures allows us to take account of the layered, multifaceted dimensions of perceptual response without falling prey to a biological, environmental, or cultural reductionism. The porosity and responsiveness of a biocultural creature to its own modes of persisting in its habitats entails, first, that we not conceive of living matter as impenetrable, and second, that we not think of the traffic of habitats into biocultural creatures as a deposit, an imprint, or a sedimentation that is determinative of what they are. For biocultural creatures, the noncontemporaneity of perceptual response and the persistence of the form, the means, and the modes of responding over time and generations mean that biocultural creatures are not identical with the habitats that culture them. The insight here is not unfamiliar, for theorists have negotiated a similar tangle in thinking about the constitution of subjects via language, discourse, or culture. Just as various forms of social theory account for the ways that identities are formed by but exceed the normative contexts of their formation, so for biocultural creatures embodied existence is ineluctably constituted through but not coincident with or reducible to current habitats. It is critically important to acknowledge the incorporation and embedding of responses to social and material habitats because in doing so we account for humans' profound dependence on our habitats, for the myriad ways that identity, affects, norms, germs, nutrition, pollution, and other forces and phenomena have their acculturating effects on our development, growth, and persistence in living. At the same time, to conceive of some of those effects as anticipatory, as having a differential temporal horizon, is to be able to distinguish between the self and habitat, to account for sparks of resistance, refusal, innovation, endurance, surprise, brilliance, judgment, decision, and so forth, in cultural and political life.

. . . Living in Biocultural Habitats

One of the important implications of the conceptualization of humans as biocultural creatures is the reconfiguration of what we consider to fall under the rubric of "culture." To accept the idea that humans are biocultural is to have to acknowledge that the worlds we inhabit and imbibe are what Davis and Morris (2007) call "biocultures" or what I would like to term "biocultural habitats." To conceive of biocultural creatures as perceptually responsive in the ways I have outlined is to suggest, along with Eduardo Kohn (2013), that for biocultural creatures meaning is indexical as well as symbolic. The "culture" that provokes transformative responses in humans conceived as biocultural creatures is not just symbolic and representational but also chemical, spatial, thermal, viral, bacteriological, nutritive, and so on. The responses of a biocultural creature to the transit of the material constituents of a habitat back and forth across the permeable boundaries of the body are activities that also compose and recompose a biocultural self. If we lean on the idea that "culture" is the conditions, practices, and processes of cultur*ing*, then to work with the notion that humans are biocultural creatures is to bring within the ambit of "culture" all the chemical, spatial, thermal, viral, bacteriological, and nutritive factors, as well as all the social, political, aesthetic, and economic practices that in combination, and sometimes at cross-purposes, provide the conditions through which biocultural humans grow into subjects. To think of humans and other creatures as biocultural is to have to think of habitats and "cultures" as biocultural too—it is to have to think of cultures and habitats together as biocultural habitats.

Of course, this is a weird, because relatively unfamiliar, way of thinking about culture. But in the same way that this book extends the work of many thinkers to denaturalize the "body" so as to show that what we have thought of as "biology" is biocultural, so we must deculturalize "culture."

This is an admittedly awkward neologism, but what I mean when I say we must deculturalize "culture" is that we must refuse the idea that culture is something that humans do or create by virtue of transcending, overcoming, or exceeding the stodgy, staid limitations of their fleshy incarnation. On the one hand, as Davis and Morris (2007) propose, it is to acknowledge that artistic, literary, musical, sculptural, and other cultural practices index the modes of perception, experience, movement, and expression and the possibilities of imagination of the biocultural creatures who create them (416). On the other hand, to deculturalize culture is to demand a fuller, richer, more expansive sense of the environments that culture human creatures. Under this aspect, to deculturalize culture is to appreciate more robustly the ways that subjective and collective experiences leave their mark in the flesh; the normative imperatives that drive contemporary discourses of biopolitics and settler colonialism, the pulsing of neoliberalism and empire in their various economic, cultural, and military manifestations, the entrenchment of racialized violence, and the reverberations of climate change become more or less transiently embodied, incorporated in ways that both frustrate and facilitate social and political processes of institutionalization and protest. In addition, to deculturalize culture is also to appreciate, as Elizabeth Wilson (2011), Stacy Alaimo (2010), and Nancy Tuana (2008) suggest, that the vexed, violent, erotic, textured, and sometimes fractured forms taken by human subjectivity or identity also develop in response to the chemicals, pharmaceuticals, pollutants, nutrients, foliage, built spaces, and so on that compose our habitats.

If we are to cleave to the ideas of biocultural creatures and biocultural habitats—and keep all the dimensions of the "culturing" of humans in our analytic sights—we must find ways to forge hybrid alliances between humanistic, sociological, and biological analyses of social and political life. Indeed, I think it is only if scholars commit themselves to proceeding, as Bruno Latour (2004) proposes, "from the complicated to the still more complicated" (191) in their research on biocultural creatures and biocultural habitats that we can challenge some of the dangers of oversimplification and reductionism that inevitably seem to make their appearance.

One of the dangers is that scholars might perceive the responsiveness of biocultural creatures to their habitats at the molecular level as license to ignore social or demographic categories—as if the science of epigenetics will deliver the neoliberal dream of a postracial social science. The impetus here is to portray racial differences as mere representations that overlay the "real, singular" body underneath. John Hibbing (2013) makes such a move in arguing against the presumption that the use of biology in political science entails a conservatively biased biological reductionism. Hibbing contends that the responsiveness of bodies to habitat provocations enables scholars to conceive of differences as "individual-level biological differences," differences whose individuality so diminishes the need to attend to "difference between one group and another" that greater interpersonal tolerance will flourish (481–82).

But of course, to elect to deemphasize group differences and social categories for the sake of the radical singularity of the biocultural individual is to ignore how politically freighted and contested norms and categories structure biocultural habitats and, as a consequence, come to be incorporated by the biocultural creatures living there. It is to sidestep the ways that the representation and perception of group differences, and the organization of social and political life in accordance with those representations and perceptions, create commonalities in the social and material habitats in which humans are cultured. In other words, as Ange-Marie Hancock (2013) points out, the turn away from "group difference" that putatively is made possible by contemporary biology ignores the ways that power structures social and political life. Troy Duster (2006) dismisses the notion that the responsiveness of biocultural creatures to their habitats means "the end of race" as a social category, for "real people are . . . living the *social experience* (of race)" (488; italics in original). In fact, recent findings in the field of epigenetics bolsters this claim that scholars need to account for the material and corporeal effects of the symbolic, representational, and political dimensions of biocultural habitats. In a review of recent research, Slavich and Cole (2013) observe that there is an increasingly large set of findings that individuals' *subjective* perceptions of the social world are more strongly associated with shifts in the ways that biocultural bodies compose and recompose than "actual social-environmental conditions themselves" (341; see also Fernald and Maruska 2012). If this is the case, then social scientists seeking to use biology in their work cannot bracket as irrelevant social and demographic categories wrought of politics. To the contrary, they must attend to the ways that the norms and experiences that constitute the subjective experience of being a self shape the psychological and physiological experience of being a biocultural creature living in this city, that neighborhood, this enclave, that embattled nation.

If we adopt the idea of biocultural creatures, we can account for political patterns in social epidemiology without having to reinstate race as a biological category. Thinking along these lines, Kuzawa and Sweet (2009) and Saldhana (2009) argue that the differentiation and stratification of humans along the lines of symbolic value attributed to phenotypical traits such as skin color mean that patterns of bioculturing occur that incorporate the experiences of such differentiation and stratification. In a study detailing just such a patterning, Ruby Mendenhall et al. (n.d.) find that stress has identifiable and patterned health effects among African American women living in urban Chicago, but the patterning is a result not of a biological thing called race but rather of the ways that living in the racialized political economy of a particularly violent urban Chicago neighborhood generates experiences of stress that become embedded or embodied in the biocultural humans living there. Dorothy Roberts (2011) and Jonathan Inda (2014) are quite right to remind us that the extant cultural and political lexicon of race reliably lends itself to the notion that the patterned incorporation of the experiences of differentiation and stratification can be indexed to some fantasmatic kind of racial biological substrate. However, the refusal to commit such a conceptual slippage does not require the analytic erasure of race altogether. Mendenhall et al.'s research pushes us to acknowledge that the observed patterns of stress among the women of this Chicago neighborhood must be attributed to the bioculturing effects of experiencing particular habitats as stressful—and it reminds us, as does Rutter (2012), that not everyone in a community or neighborhood experiences or responds to stress in identical ways.

Another danger is that, in cultivating what Meloni and Testa (2014) call an "epigenetic imagination," scholars might slip into thinking of biocultural habitats in only molecular terms, reducing them to mere "mechanisms" of molecular transformation that, through a "new epigenetic biopolitics," can be manipulated in target populations to produce specific results that align with elite political interests (447). The idea here is that the biopolitical management of life that scholars like Nikolas Rose (2007) trace in clinical and pharmacological settings could be expanded to micro-manage particular communities and neighborhoods—or, following

Mbembe's (2003) argument that biopolitics in some populations manifests as necropolitics, to produce exaggerated, literally noxious forms of redlining. Equally destructive, as Sarah Richardson (2015) and Richardson et al. (2014) argue, a focus on the molecular constituents of microenvironments could concentrate further regulatory control over women's bodies, their ability to birth children coding them as potent vectors of risk. Such political projects will in all likelihood be attempted, so it is critical to intercede by foregrounding the ways that such a biopolitical, epigenetic-oriented manipulation would rely on an overly local and homogenous understanding of biocultural habitats. There are two points here.

The first concerns the varied temporal and spatial scales at which biocultural habitats contribute to human development and growth. The multivalent, multimodal forms of culturing that shape human biocultural creatures in any given community include, for instance, commercial and residential zoning shaped by longer histories of racial and class segregation, the effects of international financial exchanges on regional and national job markets, intergenerational ties of family and kinship, access to food and health care, persistently structured patterns of political efficacy and disenfranchisement at the local through the national levels, changing ideas about the value of gendered labor, differential state and city investments in education and community development, national and regional urban youth culture, the threats and scars of war, and so forth. This broad swathe of culturing factors means that neither the physical space of a neighborhood, a community, or a uterine environment nor the stipulated time frame of attempted control constitute the totality of the biocultural habitat that cultures members of a community. Furthermore, as Mashoodh et al. (2012) and Guerrero-Bosagna and Skinner (2009) observe, both men *and* women are biocultural creatures whose lifelong and intergenerational openness to culturing by their biocultural habitats singly and in combination contributes to the development of their offspring. The biocultural habitats that culture humans are too extensive, complex, and nonsynchronic to make them amenable to the kinds of mastery that would be necessary to produce very particular effects in specific target populations.

The second point is that because human biocultural creatures contest ideas, resist expectations, and refuse obligation and accommodation in ways that create imaginative, social, and material frictions and striations, they should not be presumed to passively absorb their habitats or placidly become what normative imperatives demand (Hemmings 2005). Biocultural habitats are not found but rather made and remade. Through their actions intentional and unintentional, collective and in aggregate, humans create the biocultural habitats that culture them. They actively engage and transform their habitats using individual and collective activity, imagination, science, technology, language, art, economics, politics, and war to do so. Accordingly, it is important to insist on tracking and making visible the modes and manners by which hegemonic and counterhegemonic political formations contribute to and texture (the experience and re-creation of) biocultural habitats. In other words, in the same way that scientific researchers might acknowledge the different spatial and temporal scales, the accretions, disruptions, and surprises, as habitats provide sustenance, obstacles, toxins, and symbionts, so must they recognize that representational and symbolic forms of culture, experience, and meaning are temporally textured, layered, constrained, and enabled by the tensions and torsions of scale, location, and history. In pushing against the possibilities that the sciences of epigenetics will serve only to enhance biopolitical management of populations, not only do we need to consider a biocultural habitat as larger, spatially and temporally, than a neighborhood but also we need to take account of the dissent, fracture, tension, and contestation that ineluctably compose biocultural habitats.

Equally important is the related need to persist in examining how biocultural habitats inflect research about the effects of biocultural habitats on humans and other biocultural creatures. For example, as Guthman and Mansfield (2015), Ernie Hood (2005), and Celia Roberts (2003) point out, researchers tend to rely on the normatively restrictive binary understanding of sex when they investigate the effects of environmental pollutants on the development of endocrine and reproductive hormones and sexual morphology. The risk here is that a lack of critical self-reflexivity about the dominance in our biocultural habitats of assumptions that there are naturally just two sexes—when researchers like Anne Fausto-Sterling (2000) have found that bodies develop into many more than just two sexual types—could generate an epidemiological link between endocrine-disrupting chemicals in the environment and intersexuality and transgender identity formation. Such an explanatory link would position intersexuals and trans people within an obnoxious clinical framework as "problems" whose future occurrence could be prevented by environmental engineering. To point out this troubling possibility is not to say that we should not be concerned about pollutants; it is to draw attention to the ways social and political assumptions that thread through biocultural habitats contribute to the framing of research problems and questions.

In the push to resist both the clinical individualization and the abstraction of population that is characteristic of biopolitics, the idea of biocultural habitats points us to what we hold in common. The idea of biocultural habitats pushes scholars to rearticulate their concepts of human cultures and material environments so that they coincide in a common notion of a medium within which living creatures are cultured. In doing so, this idea provides an opportunity to shift the balance from administrative to democratic forms of politics, focusing our political and theoretical attention on the political project of transforming the conditions of our collective existence.

In Anticipation

The effort to elaborate on the proposal that humans are biocultural creatures may—and I hope does—provide important insights or points of departure for thinking about the configuration of political challenges confronting us and for the mapping of possibilities to meet them. But those configurations and maps can be seen as compelling only through the work of persuasion, manipulation, organization, and contested and sedimented institutionalization.

If the variants of posthumanism can be understood to undermine the conceit of human exceptionalism in its many forms, then by extension they can also be understood as finally having debunked the late Renaissance–early modern notion of the great chain of being in which the "what" of creatures is aligned with their moral value and their political dignity and authority. In this light, posthumanism is an effort to shake out the remnants of the great chain of being as it has formed our political and philosophical self-understanding. I am not sure whether this upset of a vertical ontology, in which some things simply are valued more than others, requires that we think in terms of a horizontal one—perhaps there is no plane of intrinsic value whatsoever. But what is clear is that in the resurgence of interest in ontology that marks so much of contemporary posthumanism, the last links between an early modern theistic ontology and theories of politics have been severed. Although it might be tempting to try to attach an alternative ontology to political worlds, political hierarchies, or visions of collective action, I want to caution against it. Such an impulse seems to me too redolent of the logic at the heart of the ideas of the human that have been dismantled through varieties of posthumanism.

Even if we rewrite or re-vision an ontology for the moment of posthumanism, politics conceived broadly as an effort to organize collective life cannot be thought as a necessary logical or conceptual entailment of an ontological claim about what humans are. Instead, as Stephen White (2000) suggests, a proposal that humans are biocultural creatures might, "at most, prefigure practical insight or judgment, in the sense of providing broad cognitive and affective orientation" (11). To deny that politics is a necessary entailment of ontology is not also to deny that what we (think we) are might shape what we consider to be politically important or politically expedient. Nor is it surreptitiously to revitalize a voluntarist, self-sovereign human political subject and thereby to sideline, again, the dependence of humans and human action on other humans, other biocultural creatures, and manifold dimensions of their biocultural habitats. Rather, to deny that the shape and course of politics can be spun logically out of an ontological claim is to propose that thinking well about what humans are will not obviate the need to think vigorously, collaboratively, in contestation and protestation, about what we should do.

The activities of biocultural creatures create and recreate the biocultural habitats that culture them. The idea of biocultural creatures makes it impossible to think of humans without also thinking about the biocultural habitats that culture them. To pose the question "How should we live?"—as Rose (2013: 23) suggests we must—is thus also to pose questions about the kinds of worlds we want to live in, the kinds of biocultural habitats we want to create for ourselves and others, and the kinds of creatures we intend or hope to become. Even though an argument that we should conceive of humans as biocultural creatures cannot provide a set of ethical and political prescriptions, it may, on consideration, help us find our ethical and political bearings. It is my hope that the insights developed in this book can serve in such an orienting capacity.

Notes

Chapter 1. Carbon

1 Leon Lederman and Christopher Hill (2011) point out that the unpredictability of quanta puts them at odds with the models of cause and effect that predominate in science (53). And whether the insights of quantum physics can in fact be used is currently quite contentious: some neuroscientists claim that the theory that our sense of smell depends on a match between smell receptors in the nose and the shape of odorific molecules should be replaced by a theory in which the vibration of a molecule effects a quantum shift in smell receptors such that we recognize a particular scent (Franco et al. 2011).

2 For a good explanation of the variety of subatomic particles and the forces that hold them together and break them apart, see Slaven (1998).

3 For a detailed explanation of contemporary versions of the transmutation of the elements (i.e., nuclear reactions), see Atkins and Jones (2010: 713). For a historiography, see Principe (2011).

4 Carbon 14, or heavy carbon, has a half-life of 5,730 years. This means that for a given portion of heavy carbon, it will take that long for half of it to have decayed into carbon 12 (Atkins and Jones 2010: 719).

5 If that were to happen, there would be an enormous nuclear explosion. If a single electron were to be forced into the nucleus—a process called electron

capture—then a nuclear reaction would unfold in which a proton was transformed into a neutron (Atkins and Jones 2010: 709).
6 The region or field in which an electron spins is often represented via a wave function, which is a formula I cannot explain (see Lederman and Hill 2011). But what these fields look like were we to visualize them are three-dimensional, roughly rounded, fuzzy-edged clouds—spherical, doughnut- or volcano-shaped, kind of blobby. I know: poetic language.
7 And they appear to be a tip-to-tip tear shape if seen two-dimensionally, as in a graph on a page.

Chapter 2. Membranes
1 For an opportunity to observe the vibration of molecular structures under extreme magnification, see Jmol, an open-source Java viewer for chemical structures in 3D, available at http://www.jmol.org/.
2 Interestingly, when we talk about diffusion in a liquid, we often do so with an aqueous solution in mind: diffusion takes place in water, and since the composition of our bodies is roughly 70 percent water, diffusion in aqueous solution is a dominant bodily theme. But what about water itself? Water is not simply a passive medium through which various concentrations of solutes redistribute themselves. Water also diffuses, which is to say that its molecules "tend to move from a region of higher water concentration to one of lower water concentration" (Nelson and Cox 2008: 51). However, because water is often the solution in which other solutes diffuse, the action by which water redistributes its concentration is called something else: osmosis. Osmosis is essentially diffusion from the perspective of water, and it is important because the action of osmosis can dilute the concentration of a substance in solution at the same time that that substance diffuses (Lodish et al. 2008: 444). In other words, the reduction of a concentration gradient can occur in two directions.
3 Incidentally, when all of the carbons in the chain are bonded with single covalent bonds and each holds hands with two hydrogen atoms, the fatty acid is considered saturated with hydrogens: a saturated fat. And yes, this means that if there are carbons in the chain held together with double covalent bonds—and thereby each holding hands with just one hydrogen atom rather than two—it is an unsaturated fat. And if there are more than one such double-bonding sites, it is a polyunsaturated fat . . .
4 Nelson and Cox (2008) provide a more official statement of this point, explaining that "the strength of hydrophobic interactions is not due to any intrinsic attraction between the nonpolar moieties. Rather, it results from the system's achieving the greatest thermodynamic stability by minimizing the number of ordered water molecules required to surround hydrophobic portions of the solute molecules" (49).
5 This conceptualization of cellular distinction also gives us a further reason to appreciate the work of theorists who criticize the idea that immunity concerns a

me/not me dynamic. See Haraway (1989), Pradeu and Carosella (2006), Cohen (2009), and Esposito (2011).

6 Exactly which kinds of organisms and creatures can be said to perceive is quite controversial. See Myra Hird's (2009a) discussion of bacterial semiosis and Uexküll (2010) for a discussion of creaturely perception. For an overview of debates in the field of zoosemiosis, see Barbieri (2009) and Kull et al. (2009).

7 This gate or transporter is called sodium-potassium ATP-ase, the latter part of the name indicating that it requires energy in the form of ATP in order to work. I will explain more about ATP in chapter 4.

8 There are actually negatively charged chloride ions involved in this dynamic too. They effect a balance in the electrical charges to just such an extent that the chemical gradient at rest can work. However, to add that activity into this account would make it more complicated than helpful. For a cogent explanation, see Nichols et al. (2001).

9 There is a particular organ in the brain that functions as the crossroads between the different nervous systems. Becker et al. (2002) state that in those creatures with a spine that surrounds many of the nerves, "the hypothalamus is the neural control center for all endocrine systems" (23).

10 Becker et al. (2002) point out that the regulation of cholesterol is important because "steroid hormones are all synthesized from the common precursor cholesterol" (30).

11 For example, as I have shown, transporters are critical to the production and maintenance of the electrochemical gradients the cell needs to facilitate necessary chemical reactions. Scientists have found that "transporters constitute a significant fraction of all proteins encoded in the genomes of both simple and complex organisms. There are probably a thousand or more different transporters in the human genome" (Nelson and Cox 2008: 391).

Chapter 3. Proteins

1 The question of in what respect we can describe biological activity as purposive or as driven by a teleological impulse has a long and checkered history. See, for a good overview, Wattles (2006). For a now-classic position, see Nagel (1977).

2 It turns out that it is quite difficult to do because we tend to think about causes as forces that endure, as something akin to an agent, if not a form of agency. Speculating, I want to say that this difficulty is linked to the fact that in the Latin from which we derive the notion of cause, *causare* denotes a speaker who advances a particular argument with the hope of effecting change. So perhaps, as the Greek notion of *telos* gets translated into Latin, we end up with a metaphysical syntax that keeps implying causes as persistent agents.

3 I do say "just about" here. There are one or two amino acids with a slightly different structure, but that difference is irrelevant—and probably confusing—here.

4 It is critical that proteins fold in the manner specific to their sequence; otherwise they become mutant or nonfunctioning proteins, which will be to the detriment of the cell and indeed the larger organism. Luckily, Klug et al. explain, there are proteins called chaperones that assist in the folding, by positioning pieces of the peptide string and providing energetic nudges in such a manner that the necessary folds occur in the right places and order (2009: 400).

5 The latter is the case for the proteins in mammalian muscle cells—just ten days! Keep up the exercise!

6 These signaling molecules are called methyl groups, acetyl groups, and phosphoryl groups, among others—so named for the kind of chemical molecule that it is. For an extensive treatment of the activity of such epigenetic markers, see Allis et al. (2007).

7 And with another nod to the argument made above, even though, in principle, the array of amino acids floating around in the nucleus to be used in the construction of proteins is random, in practice, in process, in a living body, it is contingent, dependent on the nutrients brought in from the environment that either supply the amino acids over time or provide the ingredients for the body and its various gut bacteria to create.

Chapter 4. Oxygen

1 A phosphate group consists of a phosphorous atom surrounded by four oxygen atoms. The outer shell of a phosphorous atom needs only three electrons to get to the stable eight. So, the fact that four oxygen atoms with their waiting-to-pair electrons are bound to the phosphorous means that the molecule as a whole has more electrons than protons: three, to be exact. This is a forced and somewhat uncomfortable fit—the electrons become highly agitated or energized—but it works for the molecule as a unit because resonance redistributes the excess negative charge throughout all four of the oxygen-phosphorous bonds.

2 It is shipped around in a variety of forms. There is one molecule called adenosine (pronounced ad-en-o-seen) that can bind with either one, two, or three phosphate molecules—in which case it is called adenosine monophosphate (one), adenosine diphosphate (two), or adenosine triphosphate (three). Indeed, these three energy molecules are so important that they get their own abbreviations in the literature—AMP, ADP, and ATP; ATP is considered the gold standard. Another molecule, called guanosine (gwan-o-seen), also performs as a carrier of phosphate groups, appearing as GDP or GTP.

3 For an accessible account, see Roach (2013). See also Koeppen and Stanton (2010).

4 Perhaps not too surprisingly, these membrane channels are called glucose transporters. As Lodish et al. (2008) explain, they are composed of proteins known as GLUT 1 through GLUT 12: these twelve different kinds of glucose transporters appear in different cell types (liver, brain, etc.) and enable faster or slower transport of glucose depending on the needs of the cell type (443).

5 Mitochondria are amazing and are one of the key features distinguishing our cells from bacterial cells. Slonczewski and Foster (2009) note that scientists speculate that mitochondria were originally a form of bacteria that got absorbed into other cells and developed a symbiotic relationship with the hosts that eventually became the cells we see today (30). Lane (2005) gives a rollicking account of how weirdly amazing it is that such a productive symbiotic relationship became a central feature of the capacity of living creatures to live.

6 These enzymes are known as nicotinamide adenine dinucleotide (NADH) and flavin adenine dinucleotide (FADH2).

7 Lodish et al. (2008) explain that these attractors are often iron ions surrounded by alternating double and single covalent carbon bonds that carry and redistribute the extra electron negative charge as it passes through via resonance (495).

8 Lodish et al. (2008) point out that another possibility here is that the oxygen will receive an electron but will not manage at the same time to snag any hydrogen ions to create water. In this case, the oxygen has a negative charge, which makes it very unstable and liable to break down into other negatively charged molecules. These unstable oxygen-derived molecules easily bind with other molecules in cells in ways they are not supposed to, i.e., in ways that are damaging because they compromise the reactivity of the cells to which they attach. This kind of damage is called oxidative stress, and it is in the hopes of countering oxidative stress that we consume antioxidants (502–3).

9 Incidentally, carbon monoxide fits quite neatly and securely into the pockets on the hemoglobin as well. When it does so, it prevents oxygen molecules from taking their place (a molecular version of an obnoxious "sorry, seat taken"), which then creates a condition of generalized oxygen deprivation throughout the body—and eventually death. Needless to say, this is why it is important to install carbon monoxide detectors in your house.

Chapter 5. Time

1 For the more various kinds of reproduction undertaken by bacteria, which do not store their DNA in a nucleus, see Myra Hird's account of microbial social life in (2009b).

2 For example, there are at least fourteen different alleles used to produce the transporter proteins or gates on cell membranes that enable the neurotransmitter serotonin to enter a cell (Nakamura et al. 2000). The genes that are used to make the proteins for the transporters are similar. What differentiates the alleles is the length of a particular sequence of DNA near the gene sequence that functions as something akin to a highlighted signpost for the transcription machinery when it undertakes to dock and transcribe the genes. These highlighted sequences are called promoters, because the distinctiveness of their sequence pattern makes the associated gene an obvious target for transcription: out, loud, and proud, promoters effectively promote their associated genes. The differential size in this promoter region affects the likelihood and frequency of transcrip-

tion, with a shorter promoter decreasing transcriptional frequency and a longer promoter increasing it. Now, since the epigenetic effects of the environment and behavior affect whether transcriptional machinery can dock and transcribe, since there are so many other proteins and biomolecules involved in the production and transport of serotonin, since receptors for serotonin can be found in many different organs and tissues and can be responsive to chemicals other than serotonin (Charney and English 2012; Rudnick 2006), and since the work of serotonin on the nervous system is bound up with the effects of other neurotransmitters (Heils et al. 1996; Conway at al. 2010) as well as broader dietary, social, and familial factors (Wilson 2011), it is not clear to scientists yet what effects such allelic differences have on the humans or other mammals in which they are found. But McGuffin et al. (2011) suggest that the trend of evidence is that these differences may have an effect on how a creature reacts to stress in its engagement with its habitat.

3 As Guerrero-Bosagna and his colleagues (2005) pose the question: "Is persistence in the conditions allowing the establishment of changed methylation patterns across lineages a sufficient attribute for such changes to be considered as evolutionary, or do such changes need to reach the threshold of mutation at the genomic level?" (347). See also Champagne (2011: 9).

Conclusion

1 For a similar argument about the productivity of boundaries—or borders—but at a larger and more political scale, see Mezzadra and Neilson (2013).

References

Agamben, Giorgio. 2003. *The Open: Man and Animal.* Trans. Kevin Attell. Stanford, CA: Stanford University Press.

Agamben, Giorgio. 1998. *Homo Sacer: Sovereign Power and Bare Life.* Trans. Daniel Heller-Roazen. Stanford, CA: Stanford University Press.

Ahmed, Sara. 2008. "Imaginary Prohibitions: Some Preliminary Remarks on the Founding Gestures of the 'New Materialism.'" *European Journal of Women's Studies* 15 (1): 23–39.

Aiello, Leslie C. 2010. "Five Years of *Homo Floresiensis*." *American Journal of Physical Anthropology* 142 (2): 167–69.

Alaimo, Stacy. 2012. "Sustainable This, Sustainable That: New Materialisms, Posthumanism, and Unknown Futures." *Proceedings of the Modern Language Association* 3 (127): 558–64.

Alaimo, Stacy. 2010. *Bodily Natures: Science, Environment, and the Material Self.* Bloomington: Indiana University Press.

Alcoff, Linda. 2005. *Visible Identities: Race, Gender, and the Self.* New York: Oxford University Press.

Alford, John R., Carolyn L. Funk, and John R. Hibbing. 2005. "Are Political Orientations Genetically Transmitted?" *American Political Science Review* 99 (2): 153–67.

Allis, C. Davis, Thomas Jenuwein, Danny Reinberg, and Marie-Laure Caparros, eds. 2007. *Epigenetics*. Cold Spring Harbor, NY: Cold Spring Harbor Laboratory Press.

Anslyn, Eric V., and Dennis A. Dougherty. 2006. *Modern Physical Organic Chemistry*. Sausalito, CA: University Science Books.

Åsberg, Cecelia, and Lynda Birke. 2010. "Biology Is a Feminist Issue: Interview with Lynda Birke." *European Journal of Women's Studies* 17 (4): 413–23.

Atkins, Peter, and Loretta Jones. 2010. *Chemical Principles: The Quest for Insight*. 5th ed. New York: Freeman.

Barad, Karen. 2012. "On Touching—The Inhuman That Therefore I Am." *differences: A Journal of Feminist Cultural Studies* 23 (3): 206–23.

Barad, Karen. 2007. *Meeting the Universe Halfway: Quantum Physics and the Entanglement of Matter and Meaning*. Durham, NC: Duke University Press.

Barad, Karen, and Adam Kleinman. 2012b. "Intra-Actions: Interview of Karen Barad by Adam Kleinman." *Mousse* 34 (summer): 76–81.

Barbieri, Marcello. 2009. "A Short History of Biosemiotics." *Biosemiotics* 2:221–45.

Barlowe, Denise P., and Marisa S. Bartolomei. 2007. "Genomic Imprinting in Mammals." In *Epigenetics*, edited by C. Davis Allis, Thomas Jenuwein, Danny Reinberg, and Marie-Laure Caparros. Cold Spring Harbor, NY: Cold Spring Harbor Press, 357–76.

Bateman, Andrew, Ava Singh, Thomas Kral, and Samuel Solomon. 1989. "The Immune-Hypothalamic-Pituitary-Adrenal Axis." *Endocrine Reviews* 10 (1): 92–112.

Baynton, Douglas C. 2001. "Disability and the Justification of Inequality in American History." In *The New Disability History: American Perspectives*, edited by Paul K. Longmore and Lauri Umansky. New York: New York University Press, 33–57.

Becker, Jill B., S. Marc Breedlove, David Crews, and Margaret M. McCarthy. 2002. *Behavioral Endocrinology*. 2nd ed. Cambridge, MA: MIT Press.

Beckwith, Jon, and Corey Morris. 2008. "Twin Studies of Political Behavior: Untenable Assumptions?" *Perspectives on Politics* 6 (4): 785–91.

Bejerano, Gill, Michael Pheasant, Igor Makunin, Stuart Stephen, W. James Kent, John S. Mattick, and David Haussler. 2004. "Ultraconserved Elements in the Human Genome." *Science* 304 (May): 1321–25.

Bennett, Jane. 2010. *Vibrant Matter: A Political Ecology of Things*. Durham, NC: Duke University Press.

Birke, Lynda. 2000. *Feminism and the Biological Body*. New Brunswick, NJ: Rutgers University Press.

Bordo, Susan. 1987. *The Flight to Objectivity: Essays on Cartesianism and Culture*. Albany: State University of New York Press.

Boyce, W. Thomas, Marla B. Sokolowski, and Gene E. Robinson. 2012. "Toward a New Biology of Social Adversity." *Proceedings of the National Academy of Sciences* 109 (October 16): 17143–48.

Braidotti, Rosi. 2013. *The Posthuman*. Malden, MA: Polity.

Brooker, Robert J., Eric P. Widmaier, Linda Graham, and Peter Stiling. 2008. Selected Material from Biology. Vol. 1. Chemistry, Cell Biology and Genetics. New York: McGraw Hill.

Brown, Wendy. 2005. Edgework: Critical Essays on Knowledge and Politics. Princeton, NJ: Princeton University Press.

Brown, Wendy. 2004. "'The Most We Can Hope For . . .': Human Rights and the Politics of Fatalism." South Atlantic Quarterly 103 (2/3): 451–63.

Brown, Wendy. 2001. Politics out of History. Princeton, NJ: Princeton University Press.

Bryant, Levi. 2011. The Democracy of Objects. Ann Arbor, MI: Open Humanities Press.

Bunch, Charlotte, and Samantha Frost. 2000. "Women's Human Rights." In Routledge International Encyclopedia of Women: Global Women's Issues and Knowledge, edited by Cheris Kramarae and Dale Spender. New York: Routledge, 1078–83.

Bush, Guy L. 1994. "Sympatric Speciation in Animals: New Wine in Old Bottles." Trends in Ecology and Evolution 9 (8): 285–88.

Bush, Guy L. 1969. "Sympatric Host Race Formation and Speciation in Frugivorous Flies of the Genus Rhagoletis (Diptera, Tephritidae)." Evolution 23 (June): 237–51.

Butler, Judith. 2010. Frames of War: When Is Life Grievable? New York: Verso.

Butler, Judith. 2006. Precarious life: The Powers of Mourning and Violence. New York: Verso.

Butler, Judith. 2004. Undoing Gender. New York: Routledge.

Butler, Judith. 1993. Bodies That Matter: On the Discursive Limits of "Sex." New York: Routledge.

Callaway, Ewen. 2014. "Modern Human Genomes Reveal Our Inner Neanderthal." Nature. doi:10.1038/nature.2014.14615.

Canguilhem, Georges. 2008. Knowledge of Life. Trans. Stefanos Geroulanos and Daniela Ginsburg. New York: Fordham University Press.

Castoriadis, Cornelius. 1987. The Imaginary Institution of Society. Trans. Kathleen Blamey. Cambridge, MA: MIT Press.

Cavell, Stanley. 1979. The Claim of Reason: Wittgenstein, Skepticism, Morality, and Tragedy. New York: Oxford University Press.

Cavell, Stanley, Cora Diamon, John McDowell, Ian Hacking, and Cary Wolfe. 2008. Philosophy and Animal Life. New York: Columbia University Press.

Chakrabarty, Dipesh. 2009. "Climate of History: Four Theses." Critical Inquiry 35 (winter): 197–222.

Champagne, Frances A. 2011. "Maternal Imprints and the Origins of Variation." Hormones and Behavior 60:4–11.

Chan, William F. N., Cecile Curnot, Thomas J. Montine, Joshua A. Sonnen, Katherine A. Guthrie, and J. Lee Nelson. 2012. "Male Microchimerism in the Human Female Brain." PLoS One 7 (9): e45592. doi:1371/journal.pone.0045592.

Charney, Evan. 2012. "Behavior Genetics and Post Genomics." Behavioral and Brain Sciences 35 (5): 331–58.

Charney, Evan. 2008. "Genes and Ideologies." Perspectives on Politics 6 (2): 321–28.

Charney, Evan, and William English. 2012. "Candidate Genes and Political Behavior." *American Political Science Review* 106 (1): 1–34.

Cheah, Pheng. 1996. "Mattering." *Diacritics* 26 (1) (spring): 108–39.

Chen, Mel. 2012. *Animacies: Biopolitics, Racial Mattering, and Queer Affect.* Durham, NC: Duke University Press.

Code, Lorraine. 2006. *Ecological Thinking: Thinking the Politics of Epistemic Location.* Oxford: Oxford University Press.

Cohen, Ed. 2009. *A Body Worth Defending: Immunity, Biopolitics, and the Apotheosis of the Modern Body.* Durham, NC: Duke University Press.

Cole, Steve W. 2009. "Social Regulation of Human Gene Expression." *Current Directions in Psychological Science* 18 (3): 132–37.

Collins, Patricia Hill. 1998. "It's All In the Family: Intersections of Gender, Race, and Nation." *Hypatia* 13 (3): 62–82.

Connolly, William. 2013a. *The Fragility of Things: Self-Organizing Processes, Neoliberal Fantasies, and Democratic Activism.* Durham, NC: Duke University Press.

Connolly, William. 2013b. "The 'New Materialism' and the Fragility of Things." *Millennium—Journal of International Studies* 41 (3): 399–412.

Connolly, William. 2011. *A World of Becoming.* Durham, NC: Duke University Press.

Connolly, William. 2002. *Neuropolitics: Thinking, Culture, Speed.* Minneapolis: University of Minnesota Press.

Connolly, William. 1999. *Why I Am Not a Secularist.* Minneapolis: University of Minnesota Press.

Conway, Christopher C., Constance Hammen, Patricia A. Brennan, Penelope A. Lind, and Jake M. Najman. 2010. "Interaction of Chronic Stress with Serotonin Transporter and Catechol-O-Methyltransferase Polymorphisms in Predicting Youth Depression." *Depression and Anxiety* 27 (8): 737–45.

Coole, Diana. 2013. "Agentic Capacities and Capacious Historical Materialism: Thinking with New Materialisms in the Political Sciences." *Millennium—Journal of International Studies* 41 (3): 451–69.

Coole, Diana. 2005. "Rethinking Agency: A Phenomenological Approach to Embodiment and Agentic Capacities." *Political Studies* 53:124–42.

Cooper, A., and C. B. Stringer. 2013. "Did the Denisovan Cross Wallace's Line?" *Science* 342 (18 October): 321–23.

Coyne, Jerry A. "Ernst Mayr and the Origin of Species." *Evolution* 48 (1): 19–30.

Crutzen, Paul J., and Eugene F. Stoermer. 2000. "The 'Anthropocene.'" *International Geosphere-Biosphere Programme (IGBP) Newsletter* 41 (May): 17–18.

Daniels, Cynthia. 2006. *Exposing Men: The Science and Politics of Male Reproduction.* New York: Oxford University Press.

Daniels, Cynthia. 1997. "Between Fathers and Fetuses: The Social Construction of Male Reproduction and the Politics of Fetal Harm." *SIGNS: Journal of Women in Culture and Society* 22 (3): 579–616.

Davis, Lennard J., and David B. Morris. 2007. "Biocultures Manifesto." *New Literary History* 38 (3) (summer): 411–18.

Davis, Noela. 2009. "New Materialism and Feminism's Anti-Biologism: A Response to Sara Ahmed." *European Journal of Women's Studies* 16 (1): 67–80.

Dawkins, Richard. 1976. *The Selfish Gene*. New York: Oxford University Press.

Delanda, Manuel. 1997. *A Thousand Years of Nonlinear History*. Brooklyn: Zone Books.

Dennett, Daniel C. 1996. *Darwin's Dangerous Idea: Evolution and the Meanings of Life*. New York: Touchstone.

Duster, Troy. 2006. "Lessons from History: Why Race and Ethnicity Have Played a Major Role in Biomedical Research." *Journal of Law, Medicine and Ethics* 34 (3): 487–96.

Duster, Troy. 2005. "Race and Reification in Science." *Science* 307 (18 February): 1050–51.

Esposito, Roberto. 2012. *Third Person: Politics of Life and Philosophy of the Impersonal*. Malden, MA: Polity.

Esposito, Roberto. 2011. *Immunitas: The Protection and Negation of Life*. Malden, MA: Polity.

Esposito, Roberto. 2009. *Communitas: The Origin and Destiny of Community*. Trans. Timothy Campbell. Stanford, CA: Stanford University Press.

Esposito, Roberto. 2008. *Bios: Biopolitics and Philosophy*. Minneapolis: University of Minnesota Press.

Fausto-Sterling, Anne. 2012. *Sex/Gender: Biology in a Social World*. New York: Routledge.

Fausto-Sterling, Anne. 2008. "The Bare Bones of Race." *Social Studies of Science* 38 (5): 657–94.

Fausto-Sterling, Anne. 2005. "The Bare Bones of Sex: Part 1—Sex and Gender." *SIGNS: Journal of Women in Culture and Society* 30 (2): 1491–527.

Fausto-Sterling, Anne. 2004. "Refashioning Race: DNA and the Politics of Health Care." *differences: A Journal of Feminist Cultural Studies* 15 (3): 1–37.

Fausto-Sterling, Anne. 2000. *Sexing the Body: Gender Politics and the Construction of Sexuality*. New York: Basic Books.

Fernald, Russell D., and Karen P. Maruska. 2012. "Social Information Changes the Brain." *Proceedings of the National Academy of Sciences* 109 (October 16): 17194–99.

Foote, Stephanie, and Elizabeth Mazzolini, eds. 2012. *Histories of the Dustheap: Waste, Material Cultures, Social Justice*. Cambridge, MA: MIT Press.

Foucault, Michel. 2009. *Security, Territory, Population: Lectures at the College de France 1977–1978*. Trans. Graham Burchell. New York: Picador.

Foucault, Michel. 2003. *"Society Must Be Defended": Lectures at the College de France 1975–1976*. Trans. David Macey. New York: Picador.

Foucault, Michel. 2000. *Power: Essential Works of Foucault 1954–1984*. Vol. 3. Ed. James D. Faubion. Trans. Robert Hurley et al. New York: New Press.

Foucault, Michel. 1978. *The History of Sexuality*. Vol. 1. *An Introduction*. New York: Vintage.

Fowler, James H., and Darren Schrieber. 2008. "Biology, Politics, and the Emerging Science of Human Nature." *Science* 322 (7 November): 912–14.

Franco, Maria Isabel, Luca Turin, Andreas Mershin, and Efthimios M. C. Skoulakis. 2011. "Molecular Vibration-Sensing Component in Drosophilia Melanogaster Olfaction." *Proceedings of the National Academy of the Sciences* 108 (9) (1 March): 3797–802.

Franklin, Sarah. 2000. "Life Itself: Global Nature and the Genetic Imaginary." In *Global Nature, Global Culture*, edited by Sarah Franklin, Celia Lury, and Jackie Stacey. London: Sage Press, 188–227.

Frost, Samantha. 2014. "Reconsidering the Turn to Biology in Feminist Theory." *Feminist Theory* 15 (3): 307–26.

Frost, Samantha. 2008. *Lessons from a Materialist Thinker: Hobbesian Reflections on Ethics and Politics*. Stanford, CA: Stanford University Press.

Fullwiley, Duana. 2015. "Race, genes, power." *British Journal of Sociology* 66 (1): 36–45.

Fynsk, Christopher. 2004. *The Claim of Language: A Case for the Humanities*. Minneapolis: University of Minnesota Press.

Garrett, Reginald H., and Charles M. Grisham. 2010. *Biochemistry*. 4th ed. Boston: Brooks/Cole.

Goodman, Alan H., and Thomas L. Leatherman. 1998. *Building a New Biocultural Synthesis: Political-Economic Perspectives on Human Biology*. Ann Arbor: University of Michigan Press.

Grosz, Elizabeth. 2011. *Becoming Undone: Darwinian Reflections on Life, Politics, and Art*. Durham, NC: Duke University Press.

Grosz, Elizabeth. 2004. *The Nick of Time: Politics, Evolution, and the Untimely*. Durham, NC: Duke University Press.

Guerrero-Bosagna, Carlos, Pablo Sabat, and Luis Valladares. 2005. "Environmental Signaling and Evolutionary Change: Can Exposure of Pregnant Mammals to Environmental Estrogens Lead to Epigenetically Induced Evolutionary Changes in Embryos?" *Evolution and Development* 7 (4): 341–50.

Guerrero-Bosagna, Carlos M., and Michael K. Skinner. 2009. "Epigenetic Transgenerational Effects of Endocrine Disruptors on Male Reproduction." *Seminar on Reproductive Medicine* 27 (5) (September): 403–8.

Gunnarsson, Lena. 2013. "The Naturalistic Turn in Feminist Theory: A Marxist-Realist Contribution." *Feminist Theory* 14 (1): 3–19.

Guthman, Julie, and Becky Mansfield. 2015. "Plastic People." *Aeon* (23 February). Accessed 15 September 2015. http://aeon.co/magazine/science/have-we-drawn-the-wrong-lessons-from-epigenetics/.

Guthman, Julie, and Becky Mansfield. 2013. "The Implications of Environmental Epigenetics: A New Direction for Geographic Inquiry on Health, Space, and Nature-Society Relations." *Progress in Human Geography* 37 (4): 486–504.

Habermas, Jürgen. 2003. *The Future of Human Nature*. Malden, MA: Polity Press.

Hall, Judith G. 1988. "Review and Hypotheses: Somatic Mosaicism: Observations Related to Clinical Genetics." *American Journal of Human Genetics* 43:355–63.

Hammonds, Evelynn M., and Rebecca M. Herzig. 2008. *The Nature of Difference: Sciences of Race in the United States from Jefferson to Genomics.* Cambridge, MA: MIT Press.

Hancock, Ange-Marie. 2013. "Neurobiology, Intersectionality, and Politics: Paradigm Warriors in Arms?" *Perspectives on Politics* 11 (2): 504–7.

Haraway, Donna. 2008. *When Species Meet.* Minneapolis: University of Minnesota Press.

Haraway, Donna. 1997. *Modest_Witness@Second_ Millennium.FemaleMan _Meets_ OncoMouse.* New York: Routledge.

Haraway, Donna. 1991. *Simians, Cyborgs and Women: The Reinvention of Nature.* New York: Routledge.

Haraway, Donna. 1989. "The Biopolitics of Postmodern Bodies: Determinations of Self in Immune System Discourse." *differences: A Journal of Feminist Cultural Studies* 1 (1): 3–43.

Harding, Sandra, ed. 2004. *The Feminist Standpoint Theory Reader: Intellectual and Political Controversies.* New York: Routledge.

Harman, Graham. 2011. *The Quadruple Object.* Hants, England: Zero Books.

Hatemi, Peter, and Rose McDermott. 2011. "The Normative Implications of Biological Research." *PS: Political Science and Politics* (April): 325–29.

Hatemi, Peter, et al. 2009. "Genetic Influences on Social Attitudes over the Life Course." *Journal of Politics* 71 (3): 1141–56.

Hayles, N. Katherine. 1999. *How We Became Posthuman: Virtual Bodies in Cybernetics, Literature, and Informatics.* Chicago: University of Chicago Press.

Heils, Armin, Andreas Teufel, Susanne Petri, Gerald Stöber, Peter Riederer, Dietmar Gengel, and K. Peter Lesch. 1996. "Allelic Variation of Human Serotonin Transporter Gene Expression." *Journal of Neurochemistry* 66 (6): 2621–24.

Hemmings, Clare. 2005. "Invoking Affect: Cultural Theory and the Ontological Turn." *Cultural Studies* 19 (5): 548–67.

Hibbing, John. 2013. "Ten Misconceptions Concerning Neurobiology and Politics." *Perspectives on Politics* 11 (2): 475–89.

Hird, Myra. 2009a. "Feminist Engagements with Matter." *Feminist Studies* 35 (2): 329–46.

Hird, Myra. 2009b. *The Origins of Sociable Life: Evolution after Science Studies.* New York: Palgrave Macmillan.

Hird, Myra. 2004. "Feminist Matters: New Materialist Considerations of Sexual Difference." *Feminist Theory* 5 (2): 223–32.

Hird, Myra. 2000. "Gender's Nature: Intersexuals, Transsexuals, and the 'Sex'/'Gender' Binary." *Feminist Theory* 1 (3): 347–64.

Hirschmann, Nancy. 2013. "Queer/Fear: Disability, Sexuality, and the Other." *Journal of the Medical Humanities* 34:139–47.

Hood, Ernie. 2005. "Are EDCs Blurring Issues of Gender?" *Environmental Health Perspectives* 113 (10): A670–A677.

Hubbard, Ruth. 1990. *The Politics of Women's Biology*. New Brunswick, NJ: Rutgers University Press.

Inda, Jonathan Xavier. 2014. *Racial Prescriptions: Pharmaceuticals, Difference, and the Politics of Life*. Burlington, VT: Ashgate.

Irigaray, Luce. 1985. *Speculum of the Other Woman*. Trans. Gillian Gill. Ithaca, NY: Cornell University Press.

Jablonka, Eva, and Marion Lamb. 2005. *Evolution in Four Dimensions: Genetic, Epigenetic, Behavioral, and Symbolic Variation in the History of Life*. Cambridge, MA: MIT Press.

Jiggins, Chris D. 2006. "Sympatric Speciation: Why the Controversy?" *Current Biology* 16 (9): 333–34. doi: 10.1016/j.cub.2006.03.077.

Jirtle, Randy L., Miriam Sander, and J. Carl Barrett. 2000. "Genomic Imprinting and Environmental Disease Susceptibility." *Environmental Health Perspectives* 108 (3): 271–78.

Jones, Maitland, Jr., and Steven A. Fleming. 2010. *Organic Chemistry*. 4th ed. New York: Norton.

Kandler, Christian, Wiebke Bleidorn, and Rainer Riemann. 2012. "Left or Right? Sources of Political Orientation: The Roles of Genetic Factors, Cultural Transmission, Assortative Mating, and Personality." *Journal of Personality and Social Psychology* 102 (3): 633–45.

Keller, Evelyn Fox. 2010. *The Mirage of a Space between Nature and Nurture*. Durham, NC: Duke University Press.

Keller, Evelyn Fox. 2000. *The Century of the Gene*. Cambridge, MA: Harvard University Press.

Kirby, Vicki. 1997. *Telling Flesh: The Substance of the Corporeal*. New York: Routledge.

Kirby, Vicki, and Elizabeth Wilson. 2011. "Feminist Conversations with Vicki Kirby and Elizabeth A. Wilson." *Feminist Theory* 12 (2): 227–34.

Klug, William S., Michael R. Cummings, Charlotte A. Spencer, and Michael A. Palladino. 2009. *Concepts of Genetics*. 9th ed. New York: Pearson Education.

Koeppen, Bruce M., and Bruce A. Stanton, eds. 2010. *Berne and Levy Physiology*. 6th ed. Philadelphia: Mosby Elsevier.

Kohn, Eduardo. 2013. *How Forests Think: Toward an Anthropology beyond the Human*. Berkeley: University of California Press.

Krause, Sharon. 2011. "Bodies in Action: Corporeal Agency and Democratic Politics." *Political Theory* 39 (3): 299–324.

Kristensen, Anders R., Joerg Gsponer, and Leonard J. Foster. 2013. "Protein Synthesis Rate Is the Predominant Regulator of Protein Expression during Differentiation." *Molecular Systems Biology* 9:689.

Kull, Kalevi, Terrence Deacon, Claus Emmeche, Jesper Hoffmeyer, and Frederick Stjernfelt. 2009. "Theses on Biosemiotics: Prolegomena to a Theoretical Biology." *Biological Theory* 4 (2): 167–73.

Kültz, Dietmar, David F. Clayton, Gene E. Robinson, Craig Albertson, Hannah V. Carey, Molly E. Cummings, Ken Dewar, Scott V. Edwards, Hans A. Hofmann, Louis J. Gross, Joel G. Kingsolver, Michael J. Meaney, Barney A. Schlinger, Alexander W. Shingleton, Marla B. Sokolowski, George N. Somero, Daniel C. Stanzione, and Anne E. Todgham. 2013. "New Frontiers for Organismal Biology." *BioScience* 63 (6): 464–71.

Kuzawa, Christopher W., and Elizabeth Sweet. 2009. "Epigenetics and the embodiment of race: Developmental Origins of US racial disparities in cardiovascular health." *American Journal of Human Biology* 21 (1): 2–15.

Landecker, Hannah. 2011. "Food as exposure: Nutritional epigenetics and the new metabolism." *Biosocieties* 6 (2): 167–94.

Landecker, Hannah, and Aaron Panofsky. 2013. "From Social Structure to Gene Regulation, and Back: A Critical Introduction to Environmental Epigenetics for Sociology." *Annual Review of Sociology* 39:333–57.

Lane, Nick. 2010. *Life Ascending: The Ten Great Inventions of Evolution*. New York: Norton.

Lane, Nick. 2005. *Power, Sex, Suicide: Mitochondria and the Meaning of Life*. Oxford: Oxford University Press.

Laqueur, Thomas. 1992. *Making Sex: Body and Gender from the Greeks to Freud*. Cambridge, MA: Harvard University Press.

Latour, Bruno. 2013. *An Inquiry into Modes of Existence: An Anthropology of the Moderns*. Trans. Catherine Porter. Cambridge, MA: Harvard University Press.

Latour, Bruno. 2007. *Reassembling the Social: An Introduction to Actor-Network Theory*. New York: Oxford University Press.

Latour, Bruno. 2004. *Politics of Nature: How to Bring the Sciences into Democracy*. Cambridge, MA: Harvard University Press.

Latour, Bruno. 1993. *We Have Never Been Modern*. Trans. Catherine Porter. Cambridge, MA: Harvard University Press.

Lederman, Leon M., and Christopher T. Hill. 2011. *Quantum Physics for Poets*. Amherst, NY: Prometheus Books.

Lewontin, Richard. 2002. *The Triple Helix: Gene, Organism, and Environment*. Cambridge, MA: Harvard University Press.

Lewontin, Richard, and Richard Levins. 2007. *Biology under the Influence: Dialectical Essays on Ecology, Agriculture, and Health*. New York: Monthly Review Press.

Leys, Ruth. 2011. "The Turn to Affect: A Critique." *Critical Inquiry* 37 (spring): 434–72.

Lloyd, Genevieve. 1984. *The Man of Reason: 'Male' and 'Female' in Western Philosophy*. London: Routledge.

Lock, Margaret. 2013. "The Epigenome and Nature/Nurture Reunification: A Challenge for Anthropology." *Medical Anthropology* 32 (4):291–308.

Lodish, Harvey, Arnold Berk, Chris A. Kaiser, Monty Krieger, Matthew P. Scott, Anthony Bretscher, Hidde Ploegh, and Paul Matudaira. 2008. *Molecular Cell Biology*. 6th ed. New York: Freeman.

Longino, Helen. 2002. *The Fate of Knowledge*. Princeton, NJ: Princeton University Press.

Lordkipanidze, David, Marcia S. Ponce de Leon, Ann Margvelashvili, Yoel Rak, G. Philip Rightmire, Abesalom Vekua, and Christoph P. E. Zollikofer. 2013. "A Complete Skull from Dmanisi, Georgia, and the Evolutionary Biology of Early Homo." *Science* 342 (18 October): 326–31.

Luciano, Dana, and Mel Y. Chen. 2015. "Has the Queer Ever Been Human?" *GLQ: A Journal of Gay and Lesbian Studies* 21 (2–3): 182–2071.

Lupski, James R. 2013. "Genome Mosaicism: One Human, Multiple Genomes." *Science* 341 (July 26): 358–59.

Macpherson, C. B. 1962. *The Political Theory of Possessive Individualism: Hobbes to Locke*. New York: Oxford University Press.

Mansfield, Becky. 2012a. "Environmental Health as Biosecurity: 'Seafood Choices,' Risk, and the Pregnant Woman as Threshold." *Annals of the Association of American Geographers* 102 (5): 969–76.

Mansfield, Becky. 2012b. "Gendered Biopolitics of Public Health: Regulation and Discipline in Seafood Consumption Advisories." *Environment and Planning D: Society and Space* 30:588–602.

Marder, Michael. 2013. *Plant-Thinking: A Philosophy of Vegetal Life*. New York: Columbia University Press.

Martin, Emily. 1991. "The Egg and the Sperm: How Science Has Constructed a Romance Based on Stereotypical Male-Female Roles." *SIGNS: Journal of Women in Culture and Society* 16 (3): 485–501.

Mashoodh, Rahia, Becca Franks, James P. Curley, and Frances A. Champagne. 2012. "Paternal Social Enrichment Effects on Maternal Behavior and Offspring Growth." *Proceedings of the National Academy of Science* 109 (October 16): 17232–38.

Massumi, Brian. 2002. *Parables for the Virtual: Movement, Affect, Sensation*. Durham, NC: Duke University Press.

Mayr, Ernst. 2002. *What Evolution Is*. London: Phoenix.

Mayr, Ernst. 1999 [1942]. *Systematics and the Origin of Species from the Viewpoint of a Zoologist*. Cambridge, MA: Harvard University Press.

Mbembe, Achille. 2003. "Necropolitics." *Public Culture* 15 (1): 11–40.

McEwen, Bruce S. 2012. "Brain on Stress: How the Social Environment Gets under the Skin." *Proceedings of the National Academy of Science* 109 (October 16): 17180–85.

McEwen, Bruce S. 2001. "Estrogen's Effects on the Brain: Multiple Sites and Molecular Mechanisms." *Journal of Applied Physiology* 91:2785–801.

McGuffin, Peter, Shaza Alsabban, and Rudolf Uher. 2011. "The Truth about Genetic Variation in the Serotonin Transporter Gene and Response to Stress and Medication." *British Journal of Psychiatry* 198:424–27.

M'charek, Amade. 2010. "Fragile Differences, Relational Effects: Stories about the Materiality of Race and Sex." *European Journal of Women's Studies* 17 (4): 307–22.

Meloni, Maurizio. 2014. "Biology without Biologism: Social Theory in a Postgenomic Age." *Sociology* 48 (4): 731–46.

Meloni, Maurizio, and Giuseppe Testa. 2014. "Scrutinizing the Epigenetics Revolution." *BioSocieties* 9 (4): 431–56.

Mendenhall, Ruby, Loren Henderson, Barbara Scott, Lisa Butler, Kedir N. Turi, Lashuna Mallett, Bobbie Wren, Rochester Bailey, Andrew Greenlee, Gene E. Robinson, Brent W. Roberts, Sandra Rodriguez-Zas, James E. Brooks, and Christy Lleras. N.d. "Involving Urban Single Low-Income African American Mothers in Genomic Research: Giving Voice to How Place Matters in Health Disparities and Prevention Strategies." Unpublished paper.

Mezzadra, Sandro, and Brett Neilson. 2013. *Border as Method, or, the Multiplication of Labor*. Durham, NC: Duke University Press.

Mills, Charles. 1997. *The Racial Contract*. Ithaca, NY: Cornell University Press.

Montoya, Michael J. 2007. "Bioethnic Conscription: Genes, Race, and Mexicana/o Ethnicity in Diabetes Research." *Cultural Anthropology* 22 (1) (Feb): 94–128.

Morton, Timothy. 2013a. *Hyperobjects: Philosophy and Ecology after the End of the World*. Minneapolis: University of Minnesota Press.

Morton, Timothy. 2013b. *Realist Magic: Objects, Ontology, Causality*. Ann Arbor, MI: Open Humanities Press.

Morton, Timothy. 2012. *The Ecological Thought*. Cambridge, MA: Harvard University Press.

Morton, Timothy. 2011. "Here Comes Everything: The Promise of Object-Oriented Ontology." *Qui Parle* 19 (2): 163–90.

Murphy, Kevin G., and Stephen R. Bloom. 2006. "Gut Hormones and the Regulation of Energy Homeostasis." *Nature* 444 (14 December): 854–59.

Mychasiuk, R., A. Harker, S. Ilnytskyy, and R. Gibb. 2013. "Paternal Stress Prior to Conception Alters DNA Methylation and Behaviour of Developing Rat Offspring." *Neuroscience* 241 (25 June): 100–105.

Mychasiuk, R., S. Ilnytskyy, O. Kovalchuk, B. Kolb, and R. Gibb. 2011. "Intensity Matters: Brain, Behavior and the Epigenome of Prenatally Stressed Rats." *Neuroscience* 180 (28 April): 105–10.

Nagel, Ernest. 1977. "Goal-Directed Processes in Biology: Teleology Revisited." *Journal of Philosophy* 74 (5) (May): 261–79.

Nakamura, M., S. Ueno, A. Sano, and H. Tanabe. 2000. "The Human Serotonin Transporter Gene Linked Polymorphism (5-HTTLPR) Shows Ten Novel Allelic Variants." *Molecular Psychiatry* 5 (2): 32–38.

Nelson, David L., and Michael M. Cox. 2008. *Principles of Biochemistry*. 5th ed. New York: Freeman.

Nichols, John G., A. Robert Martin, Bruce G. Wallace, and Paul A. Fuchs. 2001. *From Neuron to Brain*. 4th ed. Sunderland, MA: Sinnauer Associates.

Niewöhner, Jörg. 2011. "Epigenetics: Embedded bodies and the molecularisation of biography and milieu." *BioSocieties* 6 (3): 279–98.

Nixon, Rob. 2011. *Slow Violence and the Environmentalism of the Poor*. Cambridge, MA: Harvard University Press.

Nosil, Patrik. 2008. "Ernst Mayr and the Integration of Geographic and Ecological Factors in Speciation." *Biological Journal of the Linnean Society* 95:26–46.

Nussbaum, Martha. 2000. *Sex and Social Justice*. New York: Oxford University Press.

O'Huallachain, Maeve, Konrad J. Karczewski, Sherman M. Weissman, Alexander Eckehart Urban, and Michael P. Snyder. 2012. "Extensive Genetic Variation in Somatic Human Tissues." *Proceedings of the National Academy of Science* 109 (44) (October 30): 18018–23.

Okin, Susan Moller. 1979. *Women in Western Political Thought*. Princeton, NJ: Princeton University Press.

Oyama, Susan. 2001. "Terms in Tension: What Do You Do When All the Good Words Are Taken." In *Cycles of Contingency: Developmental Systems and Evolution*, edited by Susan Oyama, Paul E. Griffiths, and Russell D. Gray. Cambridge MA: MIT Press, 177–94.

Oyama, Susan. 2000a. *Evolution's Eye: A System's View of the Biology-Culture Divide*. Durham, NC: Duke University Press.

Oyama, Susan. 2000b. *The Ontogeny of Information: Developmental Systems and Evolution*. Durham, NC: Duke University Press.

Oyama, Susan, Paul E. Griffiths, and Russell D. Gray, eds. 2001. *Cycles of Contingency: Developmental Systems and Evolution*. Cambridge, MA: MIT Press.

Painter, R. C., C. Osmond, P. Gluckman, M. Hanson, D. I. W. Phillips, and T. J. Roseboom. 2008. "Transgenerational Effects of Prenatal Exposure to the Dutch Famine on Neonatal Adiposity and Health in Later Life." *International Journal of Obstetrics and Gynaecology* 115 (10): 1243–49.

Painter, R. C., Tessa J. Roseboom, and Otto P. Bleker. 2005. "Prenatal Exposure to the Dutch Famine and Disease in Later Life: An Overview." *Reproductive Toxicology* 20 (3): 345–52.

Papoulias, Constantina, and Felicity Callard. 2010. "Biology's Gift: Interrogating the Turn to Affect." *Body and Society* 16 (1): 29–56.

Pateman, Carole. 1988. *The Sexual Contract*. Palo Alto, CA: Stanford University Press.

Pinto-Correia, Clara. 1997. *The Ovary of Eve: Egg and Sperm and Preformation*. Chicago: University of Chicago Press.

Potter, Elizabeth. 2001. *Gender and Boyle's Law of Gases*. Bloomington: Indiana University Press.

Pradeu, Thomas, and Edgardo D. Carosella. 2006. "The Self Model and the Conception of Biological Identity in Immunology." *Biology and Philosophy* 21:235–52.

Pratt, Oliver, Carl Gwinnutt, and Sarah Bakewell. 2008. "The Autonomic Nervous System—Basic Anatomy and Physiology." *Update in Anaesthesia* 36–39. Ac-

cessed 15 September 2015. http://e-safe-anaesthesia.org/e_library/02
/Autonomic_nervous_system_anatomy_and_physiology_Update_2008.pdf.
Principe, Lawrence. 2011. "Alchemy Restored." *Isis* 102 (2) (June): 305–12.
Rabinow, Paul, and Carlo Caduff. 2006. "Life—After Canguilhem." *Theory, Culture and Society* 23 (2/3): 329–31.
Reilly, Niamh. 2007. "Cosmopolitan Feminism and Human Rights." *Hypatia* 22 (4): 180–98.
Ribot, Jesse. 2014. "Cause and Response: Vulnerability and Climate in the Anthropocene." *Journal of Peasant Studies* 41 (5): 667–705.
Richardson, Sarah S. 2015. "Maternal Bodies in the Postgenomic Order: Gender and the Explanatory Landscape of Epigenetics." In *Postgenomics: Perspectives on Biology after the Genome*, edited by Sarah S. Richardson and Hallam Stevens. Durham, NC: Duke University Press, 210–31.
Richardson, Sarah S., Cynthia Daniels, Matthew Gillman, Janet Golden, Rebecca Kukla, Christopher Kuzawa, and Janet Rich-Edwards. 2014. "Don't Blame the Mothers." *Nature* 512 (14 August): 131–32.
Roach, Mary. 2013. *Gulp: Adventures on the Alimentary Canal*. New York: Norton.
Roberts, Celia. 2007. *Messengers of Sex: Hormones, Biomedicine, and Feminism*. Cambridge: Cambridge University Press.
Roberts, Celia. 2003. "Drowning in a Sea of Estrogens: Sex Hormones, Sexual Reproduction and Sex." *Sexualities* 6 (2): 195–213.
Roberts, Celia. 2002. "A Matter of Embodied Fact: Sex Hormones and the History of Bodies." *Feminist Theory* 3 (1): 7–26.
Roberts, Dorothy. 2011. *Fatal Invention: How Science, Politics, and Big Business Recreate Race in the Twenty-First Century*. New York: New Press.
Robinson, Gene, Russell D. Fernald, and David F. Clayton. 2008. "Genes and Social Behavior." *Science* 322 (November): 896–900.
Robinson, Gene, Christina M. Grozinger, and Charles W. Whitfield. 2005. "Sociogenomic: Social Life in Molecular Terms." *Nature* 6 (April): 257–71.
Rose, Nikolas. 2013. "The Human Sciences in a Biological Age." *Theory, Culture and Society* 30 (1): 3–34.
Rose, Nikolas. 2007. *The Politics of Life Itself: Biomedicine, Power, and Subjectivity in the Twenty-First Century*. Princeton, NJ: Princeton University Press.
Rose, Nikolas. 1998. "Life, Reason and History: Reading Georges Canguilhem Today." *Economy and Society* 27 (2/3): 154–70.
Rudnick, G. 2006. "Structure/Function Relationships in Serotonin Transporter: New Insights from the Structure of a Bacterial Transporter." In *Neurotransmitter Transporters*, edited by H. Sitte and M. Freissmuth. Berlin: Springer-Verlag, 59–73.
Rutter, Michael. 2012. "Achievements and Challenges in the Biology of Environmental Effects." *Proceedings of the National Academy of Science* 109 (October 16): 17149–53.
Saldanha, Arun. 2009. "Reontologising Race: The Machinic Geography of Phenotype." *Environment and Planning D: Society and Space* 24 (1): 9–24.

Schiebinger, Londa. 1993. *Nature's Body: Gender in the Making of Modern Science*. Boston: Beacon Press.

Shapin, Steven, and Simon Schaffer. 1989. *Leviathan and the Air-Pump: Hobbes, Boyle, and the Experimental Life*. Princeton, NJ: Princeton University Press.

Sharan, Roded, Silpa Suthram, Ryan M. Kelley, Tanja Kuhn, Scott McCuine, Peter Uetz, Taylor Sittler, Richard Karp, and Trey Ideker. 2005. "Conserved Patterns of Protein Interaction in Multiple Species." *Proceedings of the National Academy of Sciences* 102 (6): 1974–79.

Siebler, Kay. 2012. "Transgender Transitions: Sex/Gender Binaries in a Digital Age." *Journal of Gay and Lesbian Mental Health* 16:74–99.

Singer, Merrill. 1996. "Farewell to Adaptationism: Unnatural Selection and the Politics of Biology." *Medical Anthropology Quarterly* 10 (4) (December): 496–515.

Singer, Peter. 2009 (1975). *Animal Liberation: The Definitive Classic of the Animal Movement*. New York: HarperCollins.

Skinner, Michael, Carlos Guerrero-Bosagna, M. Haque, Eric Nilsson, Ramji Bhandari, and John R. McCarrey. 2013. "Environmentally Induced Transgenerational Epigenetic Reprogramming of Primordial Germ Cells and the Subsequent Germ Line." *PLoS ONE* 8 (7): e66318. doi: 10.1371/journal.pone.0066318.

Slaven, Dave. 1998. "Dave's Microcosmos."Accessed 10 September 2015. http://webs.morningside.edu/slaven/Physics/micro/

Slavich, George M., and Steven W. Cole. 2013. "The Emerging Field of Human Social Genomics." *Clinical Psychological Science* 1 (3) (July): 331–48.

Slonczewski, Joan, and John W. Foster. 2009. *Microbiology: An Evolving Science*. New York: Norton.

Steffen, Will, Jacques Grinevald, Paul Crutzen, and John McNeill. 2011. *Philosophical Transactions of the Royal Society* 369:842–67.

Stiegler, Bernard. 1998. *Technics and Time*. Vol. 1. *The Fault of Epimetheus*. Trans. Richard Beardsworth. Stanford, CA: Stanford University Press.

Stoler, Ann. 2010. *Carnal Knowledge and Imperial Power: Race and the Intimate in Colonial Rule*. Berkeley: University of California Press.

Stoler, Ann. 1995. *Race and the Education of Desire: Foucault's History of Sexuality and the Colonial Order of Things*. Durham, NC: Duke University Press.

Sunstein, Cass, and Martha Nussbaum, eds. 2005. *Animal Rights: Current Debates and New Directions*. New York: Oxford University Press.

Tallbear, Kim. 2013. *Native American DNA: Tribal Belonging and the False Promise of Genetic Science*. Minneapolis: University of Minnesota Press.

Tanriverdi, F., L. F. G. Silveira, G. S. MacColl, and P. M. G. Boulous. 2003. "The Hypothalamic-Pituitary-Gonadal Axis: Immune Function and Autoimmunity." *Journal of Endocrinology* 176:293–304.

Tazuke, Sallie I., Cordula Schulz, Lilach Gilboa, Mignon Forgarty, Anthony P. Mahowald, Antoine Guichet, Anne Ephrussi, Cricket G. Wood, Ruth Lehmann, and Margaret Fuller. 2002. "A Germline-Specific Gap Junction Protein Required for Survival of Differentiating Early Germ Cells." *Development* (129): 2529–39.

Thayer, Zaneta M., and Christopher W. Kuzawa. 2011. "Biological Memories of past environments: Epigenetic pathways to health disparities." *Epigenetics* 6 (7) (July): 798–803.
Tuana, Nancy. 2008. "Viscous Porosity: Witnessing Katrina." In *Material Feminisms*, edited by Susan Hekman and Stacy Alimo. Bloomington: Indiana University Press, 188–213.
Uexküll, Jakob von. 2010. *A Foray into the Worlds of Animals and Humans: With a Theory of Meaning*. Trans. Joseph D. O'Neil. Minneapolis: University of Minnesota Press.
Van der Tuin, Iris. 2008. "Deflationary Logic: Response to Sara Ahmed's 'Imaginary Prohibitions: Some Preliminary Remarks on the Founding Gestures of the "New Materialism."'" *European Journal of Women's Studies* 15 (4): 411–16.
Via, Sara. 2001. "Sympatric Speciation in Animals: The Ugly Duckling Grows Up." *Trends in Ecology and Evolution* 16 (7): 381–90.
Viau, V. 2002. "Functional Cross-talk between the Hypothalamic-Pituitary-Gonadal and Adrenal Axes." *Journal of Neuroendocrinology* 14:506–13.
Wattles, Jeffrey. 2006. "Teleology Past and Present." *Zygon* 41 (2) (June): 445–64.
Webster, Jeanette I., Leonardo Tonelli, and Esther M. Sternberg. 2002. "Neuroendocrine Regulation of Immunity." *Annual Review of Immunology* 20:125–63.
Weheliye, Alexander. 2014. *Habeas Viscus: Racializing Assemblages, Biopolitics, and Black Feminist Theories of the Human*. Durham, NC: Duke University Press.
Weigman, Robyn. 2010. *Object Lessons*. Durham, NC: Duke University Press.
West-Eberhard, Mary Jane. 2005a. "Developmental Plasticity and the Origin of Species Differences." *Proceedings of the National Academy of Sciences* 2 (3 May): 6543–49.
West-Eberhard, Mary Jane. 2005b. "Phenotypic Accommodation: Adaptive Innovation Due to Developmental Plasticity." *Journal of Experimental Zoology* 304B:610–18.
West-Eberhard, Mary Jane. 1998. "Evolution in the Light of Developmental and Cell Biology, and *Vice Versa*." *Proceedings of the National Academy of Sciences* 95 (July): 8417–9.
West-Eberhard, Mary Jane. 1989. "Phenotypic Plasticity and the Origins of Diversity." *Annual Review of Ecological Systems* 20:249–78.
White, Stephen. 2000. *Sustaining Affirmation: The Strengths of Weak Ontology in Political Theory*. Princeton, NJ: Princeton University Press.
White, Yvonne A. R., Dori Woods, Yasushi Takai, Osamu Ishihara, Hiroyuki Seki, and Jonathan Tilly. 2012. "Oocyte Formation by Mitotically-Active Germ Cells Purified from Ovaries of Reproductive Age Women." *Nature Medicine* 18 (3): 413–21. Accessed via NIH Public Accessed 16 December 2013.
Wilson, Elizabeth. 2011. "Neurological Entanglements: The Case of Paediatric Depressions, SSRIs and Suicidal Ideation." *Subjectivity* 4 (3): 277–97.
Wilson, Elizabeth. 2004. *Psychosomatic: Feminism and the Neurological Body*. Durham, NC: Duke University Press.

Wilson, Elizabeth. 1998. *Neural Geographies: Feminism and the Microstructures of Cognition*. New York: Routledge.
Wittgenstein, Ludwig. 1969. *On Certainty*. Ed. G. E. M. Anscombe and G. H. von Wright. Trans. Denis Paul and G. E. M. Anscombe. New York: Harper and Row.
Wolfe, Cary. 2010. *What Is Posthumanism?* Minneapolis: University of Minnesota Press.
Wolfe, Cary, ed. 2003. *Zoontologies: The Question of the Animal*. Minneapolis: University of Minnesota Press.
Wolfe, Jeremy M., Keith R. Kluender, and Dennis Levi. 2009. *Sensation and Perception*. 2nd ed. Sunderland, MA: Sinnauer Associates.
Wolinksy, Howard. 2010. "The Puzzle of Sympatry." *European Molecular Biology Organization Reports* 11 (11): 830–33.
Yan, Zhen, Nathalie C. Lambert, Katherine A. Guthrie, Allison J. Porter, Laurence S. Loubiere, Margaret N. Madeleine, Anne M. Stevens, Heidi M. Hermes, and J. Lee Nelson. 2005. "Male Microchimerism in Women without Sons: Quantitative Assessment and Correlation with Pregnancy History." *American Journal of Medicine* 188 (18): 899–906.
Zerilli, Linda. 2005. *Feminism and the Abyss of Freedom*. Chicago: University of Chicago Press.

Index

Note: Italicized numbers indicate a figure; n indicates an endnote

absorption: cellular, 70; in glycolysis, 111–13; influence of on germ cells, 128, 129, 133, 138, 145, 148, 165n5; nerve cell absorption of and response to a habitat, 70, 73, 75–76, 122, 138, 148, 156–57; of elements of a habitat by an organism, 17, 52, 70, 111–13, 118, 133
adenosine nucleotides, 93, 141, 164n2
Agamben, Giorgio, 7–8
agency: "agentic capacities" of environmental factors, 10, 11, 12, 31; attribution of to causes, 163n2; attribution of to genes, 82; attribution of to humans, 2, 3, 5, 7, 10–13, 31–32; biological, 27, 79, 83, 90, 91; habitat-embedded creature, 13; heterogeneous assemblage, 10–11; human exceptionalism, 1, 10, 13, 147, 158; nonhuman, 10–12, 31, 105, 118, 163n2 *See also* condition of possibility; constraints
Alaimo, Stacy, 67–68, 153
alchemy, 37
amino acid: alpha carbon interactions, 86; amino acid construction: analogy of tripod, 85–86; amino beginnings and acid endings, 86; binding process of multiple amino acid molecules, 86, 87; capture of by RNA, 92, 97–98; constraining or delimiting features of an, 87; correlation of amino acids to nucleotides in a gene strand, 94–95; C-terminus (carbon-oxygen), 86; folding of polypeptides, 87–89, 137; function of DNA as molecular memory (or code) for strings, 93–96, 137, 141; influence of environment on production of in an organism, 164n7; molecules contained in an, 86;

amino acid (*continued*)
N-terminus (nitrogen), 86–87; peptide and polypeptide bonds, 86; precision of folding as an influence on protein function, 87–88; residues of in a protein, 86–87; role of functional group of an, 86; role of hydrogen atom of an, 86; shapes and topography of polypeptide strings, 87–88; sites for biochemical activity on a protein, 88; transformation of a protein from, 87–88, 97–98
amphipathic molecules, 62–63, 64
animality of human beings, 3–9, 11, 13, 25, 78, 147, 151
Anslyn, Eric V., 45
Anthropocene and anthropocentrism, 2, 3, 9, 13, 25
Aristotle, 37, 81–82
aspiration, 28, 104, 129
atoms: in amino acids, 86; anions and cations, 47; atomic elements, 32–33, 49, 59; atomism and substantialism, 33–34; as basis for matter and substance, 32–33, 40, 45, 49; bonds and bonding, 45–51, 57, 59, 110, 162n3; carbon, 38–39, 49–51, 113, 140; charges and polarity of, 36, 39–40, 44, 47–48, 50, 56, 60, 113; chemically inactive, 43; chemical reactions, 40–41, 43, 45, 49, 56–57; composition of, 36–40; as conglomerations of moving, tense energy, 32–33, 35–36, 39–45, 56–57, 60; differentiation of, 36–38, 42–43; diffusion, 57, 59–63; donors and acceptors, 49; electron-proton interactions, 40–41, 43–45, 56, 60; electrons in an atom, 36, 37, 39–50, 56, 59–60, 113; energy of as basis for organic porosity, 41, 49, 51–52, 110; energy states of, 45–46, 47–48, 108; heavy, 37, 38–39; hydrogen, 48, 59–60, 86, 102, 110, 113, 140; imbalanced, 43–46, 49; inert or balanced, 43; interactions among, 40–51; ions, 47–48, 50, 60–61; methyl groups, 140; molecules, 45–51, 56, 59, 110–11, 113, 140; neutrons in an atom, 36, 38–39; nitrogen, 38; nuclear reactions, 37–38; nuclear transmutation, 38, 161n3; nucleus of an atom, 36–38, 39, 40, 41–45, 46, 47, 56; numeric matching of protons and electrons in, 39, 40–41, 43–44, 49; numeric matching of protons and neutrons in, 38; orbitals, 42–43, 45–47, 56; oxygen, 46, 59–60, 102, 113; phosphate groups, 164n1; protons in an atom, 36–38, 40–44, 46–48, 50, 56; quantum fields and theory, 34–35, 39–40, 51, 161n1; quarks in, 36, 38; radioactivity, 38; resonance, 47, 108, 164n1; self-constraint of energy that characterizes, 35, 36–37, 41–42, 45, 51; self-interaction of energy: analogy of coffee swirling in a cup, 39–40; shells, 42–43, 45, 46, 49; solar system model of atom, 35; subatomic particles, 32, 34, 161n2; table of elements, 36–37; tensions within, 36, 37, 39–45, 56–57, 60; theft or stealing of electrons by, 44, 45, 47, 48, 49; transformation of protons and neutrons, 38; transitions in energy required by organic life, 102, 113–16; vibration of, 56–57; in water molecules, 59–61
ATP: activities of, 109–10, 111; adenosine triphosphate, 164n2; amount of in an adult human body, 109–10; ATP-ase, 163n7; components of, 109, 110; generation and regeneration of ATP molecules, 110, 112–13, 114–15; in nerve cells, 109

bacteria, 25, 75, 163n6, 164n7, 165n1, 165n5
Barad, Karen, 18, 34–35, 39–40, 51
Barlowe, Denise P., 142
Bartolomei, Marisa S., 142
Bennett, Jane, 3, 10–11, 31–32
biochemical activity: as a means of defining inside and outside, 68, 76, 77, 104, 107; as a response of an organism to its environment, 119, 129–30, 137, 148, 150; as a transition of energy in an organism,

104, 116, 117, 119; dependence of on porosity of cellular membranes, 28, 64, 67–69, 75–76, 88–90, 131; heterogeneous, 82; at level of cells, 88, 91, 107, 117; precise, sequential nature of molecular, 27–28, 83–85, 94, 106

biocultural creatures: biocultural habitats, 152–59; concept of embodied human, 17–18; convergence of qualitative and scientific research substantiating concept of, 16–17; creation of their own habitats by biocultural humans, 156–57; culturing, deculturalizing, and bioculturing, 152, 153, 155–56; definition of biology in, 4–5, 16, 149, 152; definition of culture in, 4–5; distinction of from their habitats and from other biocultural creatures, 148, 151; interactive nature of habitats and activities of, 159; moral and ethical responsibilities inherent to being, 19; ontology of human experience, 19, 159; posthumanist contribution to defining, 24–25, 158–59; psychological features of produced by their habitats, 154, 156–57, 158; reconceptualizations of relationship of human beings and their habitats, 124–25, 147–48, 151–54; reconceptualizations of what "human" means, 120–21, 158–59; temporally conditioned nature of, 157; tension and dissent as inherent to and their habitats, 157; theoretical influences on idea of, 24–26

biology: authority of as knowledge, 16; biological reductionism, 153–54; biological research, 16–17, 125–26; in concept of biological creatures, 149, 152; use of by social scientists, 16, 83, 153–57

biomolecules: epigenetic, 82, 125, 139, 140, 144, 165–66n2; function of in making an organism respond to its environment, 122, 132, 135; in gene mutation, 133, 135, 137–38, 139; germ cell, 125–26, 131, 133; passage of through generations of organisms, 122–23, 126; responsive nature of, 91, 125, 129

biopolitics: biocultural approach to, 20–21, 153; case of Afro-American women in urban Chicago, 155; colonialism, 8, 153; danger in trying to manipulate or manage populations, 15–16, 20–21, 155–58; life sciences and, 13–16, 19–21, 24; race as a concern of, 9, 14–15, 21, 154–55; social sciences and, 14, 16, 83, 153–54; societal as well as physical stress as a concern of, 126–28, 133, 155, 165–66n2. *See also* politics

Birke, Lynda, 102

blood, 74, 79, 103, 112, 115–16, 125

bonds and bonding: bonding of carbons, 50–51, 59, 162n3; bonding of protein molecules, 108; covalent, 45–47, 48, 50–51, 56, 59–61, 162n3; hydrogen, 45, 48–49; ionic, 45, 47–48, 50, 56, 60–61, 86; molecular bonding, 56–57; polar covalent bonds, 48; sharing of electrons, 45, 48, 56; theft or stealing of electrons, 44, 45, 47, 48, 49

boundaries: absence of solid between organisms and their environments, 17–18, 26–27, 41, 81, 126, 152; as a distinction of activities rather than of substances, 27, 28, 41, 52; boundary failure among groups in a society, 9; cell membranes, 54, 58–59, 67. *See also* inside and outside

Boyce, W. Thomas, 148–49

Braidotti, Rosi, 1–2

breathing: connection of to eating, 103, 111; effect of cyanide molecules in interrupting process of, 115; exhaling as a disposal of carbon waste, 102; function of in bodily energy transition processes, 107; generation of ATP molecules through, 110, 115; involvement of autonomic system in, 74; oxygen as basis for ripping-and-disposal phase of breathing, 103, 113–14, 115–16; phosphorylation and, 108–9; real and imagined roles of oxygen in, 103–5, 106. *See also* cellular respiration

Brooker, Robert J., 53, 64–67, 109

INDEX | 185

Bush, Guy L., 138
Butler, Judith, 25, 120–21

Caduff, Carlo, 16–17
Callard, Felicity, 120
Canguilhem, Georges, 24
carbohydrates, 65, 102, 111–12. *See also* glucose
carbon: alpha, 85–86; in amino acid, 85–86; as basis for life, 26, 33, 38, 49, 150; bonding of, 50–51, 59, 61, 86, 162n3, 165n7; in breathing, 115–16; in carbohydrates, 102, 111, 112; Carbon-14, 161n4; carbon atoms, 38, 42, 49–51, 102, 110–11; carbon dating, 38–39; carbon dioxide, 65, 102, 115–16; carbon monoxide, 165n9; carbon-oxygen (C-terminus) atoms, 86; carbon strings, 112; covalent bonds and, 50–51, 59–60, 61; in electron transport chains, 113; energy constraints on, 49, 150; in epigenetic molecules, 140; fatty acid chains, 59, 61, 110, 111, 162n3; in glucose, 112; heavy, 38–39, 161n4; ion construction from, 50; in Krebs cycle activity, 113; in metabolized food, 102, 110; molecular shaping capacities of, 50–51, 59; in pyruvate molecules, 112, 113, 115; in RNA, 92
Cavell, Stanley, 9
cell membranes: alleles, 137, 165n2; appearance of cell membranes: analogy of ping-pong balls in bathwater, 64; aquaporin, 66; arrangement of matter into porous, 27; compositional features of, 64–67, 78; and concepts of inside and outside, 27, 54–55, 67–68, 75–76, 77, 83; co-transport gates: analogy of gumball machine mechanisms, 66; co-transport gates, 67; diffusion and, 58–59, 64–67, 71, 107; diffusion of carbon dioxide molecules through, 115; diffusion of glucose molecules through, 112; endocrine chemicals and, 74; gates, pores, channels, transporters, and bridges, 54, 65–69, 72–74, 78–79, 90, 165–66n2; germ as means through which organisms interact with their habitats, 29, 126–30, 132–34138; influence of chemical reactions on, 68–69, 71; influence of on a cell's responsiveness to its environment, 55, 69, 135, 143; influence of on a cell's responsiveness to its own activity, 55, 69, 75, 150; influence of on an organism's response to its environment, 68–71, 91, 97, 101–2, 121, 124–25, 129; lipid bilayer, 63; in management of glucose molecules, 69; as means of creating and facilitating distinct molecular activities, 27, 28, 55, 58, 67–68; of nerve cells, 70–73, 75, 107–8; osmosis, 162n2; passage of oxygen through, 106–7; phosphorylation, 108–9; plasma membrane proteins in, 67, 79; role of in perpetuating living activity, 58, 70, 83, 85, 89–90, 98, 150; selectivity in: analogy of a gumball machine, 66; selectivity in, 65, 66–67, 70, 101, 107, 119–20, 150; tension inherent to, 53; transcorporeality of, 67–68. *See also* diffusion; influx and efflux; permeability; porosity
cells: activities within, without, and on enclosing membranes as basis for cell distinction, 58, 68, 75; amphipathic molecules of, 62–63, 64; carbohydrate management by, 111–12; cell types, 96, 125, 142–43, 164n4; cellular diffusion, 64–66, 69, 75, 107, 112; cellular structure: analogy of Los Angeles, 78–79; concentration gradients, 57, 67, 72, 107–9, 112, 114–16, 162n2; and concepts of inside and outside, 27, 54–55, 67–68, 75–76, 77, 83, 104; construction and dissolution of protein molecules by, 75, 90–91; differentiation between separate, 58, 68, 75; epigenetic markers, 96–97, 125, 139, 142–43; germ, 29, 125–33, 134–37, 138–42, 144; glucose management by, 112; high-energy molecules

in, 108; influence of habitat on, 27–28; influence of proteins and biomolecules on shapes and functions of, 125; lipid bilayer membranes of, 63; meiosis, 130–31; membranes as defining feature of, 58, 68, 75; mitosis, 130–31; molecular co-transport in, 66–67; multiple roles of proteins in, 78–79; nerve, 70–74, 108–10, 165–66n2; oxygen transience through, 28, 103, 105, 106–7, 113–14, 118; phosphorylation, 107–9; porosity of as means through which habitats shape organisms, 129–30; precision of biochemical processes of, 27–28, 68–69, 84–85, 93; protein molecules of, 78–79, 89–91, 93–94, 95, 96, 97–98; reactions of, 64–65, 68–70, 89–91, 93–94, 96, 107–9, 124; somatic, 97, 125–29, 131, 135–36, 139–40, 144; somatic mosaicism, 97; transitions of energy within, 102, 103; waste product disposal by, 102–3, 105, 106–7; water in, 78

cellular respiration: as a bodily transition of energy, 110; as a metabolic activity, 103; constrained, step-by-step nature of, 117–18; importance of carbohydrates to, 111; process of, 103, 108, 115, 117; role of oxygen in, 103, 105–6, 116–17; significance of ATP molecules to, 110. *See also* breathing; metabolism

Chakrabarty, Dipesh, 2, 13

Champagne, Frances A., 133, 135, 136, 166n3

channels: function of in cell membranes, 65–66, 67, 69; glucose transporters, 164n4; in cell membranes, 54, 65, 71, 90, 112, 143; influence of endocrine chemicals on membrane, 74; in nerve cell membranes, 70–73; potassium ion, 71–72; rhythmic functioning of nerve membrane, 72–73; sodium ion, 71–72. *See also* gates; pockets; pores

Chan, William F. N., 97

charge: absence of in neutrons, 36; as basis for moving target molecules through cellular membranes, 66; as basis for propulsion of hydrogen ions through cellular membranes, 114; cations and anions, 47; depolarization, 72; electrical as basis for interaction of polypeptides, 87, 88; electrochemical gradient and, 108, 163n8; electron of carbon, 50; electron of fatty acid molecules, 61; electron of water, 60; in carbohydrates, 111; influence of on atomic and subatomic elements, 32, 37; influence of on molecular electrochemical gradient, 114; of glucose molecules, 112; of phosphate group molecules, 108–9; polar opposite of protons and electrons, 39; resonance, 47, 164n1; shift of electron in molecules, 47; shift of electron in polar covalent bonds, 48. *See also* negative charge; polarity; positive charge

Cheah, Pheng, 25

chemical reactions: balance between protons and electrons as determinants of, 40–41; at cellular level, 65, 67–70, 75, 84, 90–91, 107, 163n11; fomented by RNA, 92, 94; influence of on cell membranes, 69–71; in construction of a protein, 97–98, 141; molecular, 49, 55, 58, 64, 65, 89, 116; precision of, 84; as product of self-constraints of energy, 49, 148; as sharing, lending, or stealing of electrons between imbalanced atoms, 43, 45, 47–48, 56; as transitions in energy, 102, 104, 107; underlying an organism's response to its environment, 97

chemical syntax, 59

chemistry, 33, 37, 45, 47

chimerism, 97

cholesterol, 64, 74, 163n10

citric acid cycle: process of Krebs or, 113–16; pyruvate molecules in the, 112–13, 115. *See also* glycolysis

climate change, 2, 12–13, 153

Cole, Steven, 17, 154

concentration: changes in cellular, 67, 69, 122; concentration gradients, 57, 67, 72, 107–9, 112, 114–16, 162n2; co-transport gates as a means of changing concentration gradients, 67, 107–8; influence of chemical reactions on molecular, 64, 65, 90, 97; influence of endocrine chemicals on of ions in cells, 74; modulation of by cellular membranes, 67–72; molecular diffusion, 57–58, 107–8, 112, 115–16; of double (pi) covalent bonds, 46–47; of electrons swirling around an atomic nucleus, 40; phosphorylation as a means of changing of high-energy molecules, 108–10; sequence in adjusting cellular, 69, 89–90; water, 162n2. *See also* diffusion; gradients

condition of possibility: cell membrane porosity as a for a cell's response to its environment, 68, 75; form as a for discreteness of an organism, 52; function of one chemical reaction as for next, 84; habitat as a for an organism's ability to live, 119; limiting quality of for creatures in a habitat, 150; matter underlying a, 32; nonhuman agency underlying human agency, 11; of energy as a constraint, 51; underlying transformation of energy into matter, 53. *See also* agency; constraints

Connolly, William, 16

constraints: activities-in-response of an organism as, 122; atoms as products of constraints on energy in relation to itself, 36, 42, 49; capacity of biological processes to have direction without intention, 83–85; constrained self-related nature of energy, 26, 31, 33, 35, 45, 51, 107; direction without intention: example of proteins, 85–90; enabling quality of energy, 49, 51–52, 150; function of at level of cells: example of nerve cells, 70–73, 75; function of at level of molecules, 56, 62, 64–65, 85, 87, 88, 150; influence of environmental on presence and nature of life, 118, 145, 148, 150;

matter as energy under particular forms of, 26, 28, 33, 36–37, 40, 53, 148; as means by which energy is shaped into something distinct and discrete, 51–52. *See also* agency; condition of possibility; precision

Coole, Diana, 10–11

Cox, Michael M., 65, 67, 68, 78, 90, 109, 162n4

C-terminus, 86. *See also* N-terminus

culture: complex, self-understanding quality of, 16–21; as cultivation, 4–5; culturing and bioculturing, 152, 153, 155–56; deculturalizing, 152–53; influence of on psychological qualities of humans, 154, 156–57, 158; influence of on shaping of bodies of organisms, 19, 23, 25, 81, 124, 148; interactive nature of habitats and activities of biocultural creatures, 159; temporal conditioning of biocultural creatures through, 157

cytosine nucleotides, 93, 141

Daniels, Cynthia, 127–28

Davis, Lennard J., 152, 153

Delanda, Manuel, 31

determinism, 6, 77, 81, 84

developmental plasticity, 132–33; as a factor in cross-generational transformations, 134–36, 138–39; as a factor in alteration of DNA structure or sequence over time, 141; as a factor in creation of new species, 139, 140, 142; as a factor in evolution of organisms, 135, 144; example of stickleback fish, 133; immunity of some proteins to, 143; regulatory genes, 138. *See also* evolution

diffusion: as activity-based means of distinguishing cellular inside from outside, 77; to adjust chemical concentrations in molecules, 64; as a means of transforming matter, 58; in and by water, 162n2; as an influence on a cell's responsiveness to its environment, 69, 75; as an influence on a cell's responsiveness to its own activity, 69, 75; as a transition in energy, 55–56, 102, 107; cell membranes and, 58–59, 64, 65–67, 78, 107; co-transport gates, 67; endocrine system management of steroids and hormones through, 74; example of diffusion: analogy of perfume in a room, 57; in glycolysis, 111, 112; gradient changes and, 71, 107, 108; in Krebs cycle, 114, 115; in management of glucose molecules, 69–70; molecular vibration and, 57; movement of down a concentration gradient, 57, 107; nerve cell management of some biochemical, 70–72; osmosis, 162n2; oxygen transit to enable high energy-demanding processes, 116; role of in perpetuating living activity, 58; role of in formation of micelles, 63; selective, 65, 66, 70, 101, 107, 119, 150. See also cell membranes; concentration

digestion. See glycolysis

DNA: alleles, 137, 165–66n2; biocultural quality of, 126–29; chimerism, 97; complementary function of to RNA, 92, 97, 131–32; copying, replication, or transcription of, 93–97, 125, 130, 136–37, 140–41; deoxyribonucleic acid characteristics, 92; developmental plasticity, 132–44; difference between genes and, 94, 95; DNA code: an example using alphabet, 94–96; double-helix ladder of, 93, 141; epigenetic markers, 96–97, 125, 140, 142; function of as molecular memory for strings of amino acids, 93–95; function of noncoding, 94; gene mutation, 137–38; germ cell, 125–26, 130, 140; influence of on protein folding and actions, 95; junk DNA, 94, 95, 96; maintenance of integrity of, 93, 95; meiosis, 130; methylation patterns, 141, 166n3; methyl groups as an influence on, 140–41; mitosis, 130; myth of überbiological, 124, 126–27; nucleotides in, 93–95; pairing preferences of, 141; porosity of cells as link between organism and environment, 126–29; power of environment or habitat to influence, 98, 128–29; regulatory, 95–97, 131, 137; response of molecules and of organisms to stimuli attached to, 82; reverse transcription of, 96–97; role of in production of protein, 93, 97–98; similarities between RNA and, 93; somatic cell, 125–26, 131, 140; somatic mosaicism, 96–97; speciation, 140

Dougherty, Dennis A., 45

Duster, Troy, 154

eating: common conceptions regarding bodily food consumption, 78, 102, 104; connection of to breathing, 103, 110; generation of ATP molecules through, 110; as part of bodily energy transition processes, 110. See also food

efflux. See influx and efflux

electrons: absence of mass of, 39; activity of in polypeptide folding, 87–88; in a ground state, 42; as a self-interaction of negative energy, 39–40; Barad on, 39–40; cations and anions, 47; in chemical bonding, 45–49, 56–57; chemically balanced or inactive atoms, 43; chemically imbalanced atoms, 43–44; common conceptions regarding movements of, 35–36; covalent bonds, 45–47, 48, 56; drawing through of in Krebs cycle, 113–15; as dynamic force in molecules, 56; electron capture, 161–62n5; electron in/determinancy: analogy of coffee swirling in a cup, 40; electron switching: example of salt, 44; electron transport chains, 113–15; hydrogen bonds, 48–49; in a water molecule, 59–60; in bonds in oxygen, 46; in carbon, 49–51; in/determinacy of, 35, 39–40; in heavy carbon, 38; in oxygen, 59–60; ionic bonds, 47–48, 56; nuclear reaction, 161–62n5; orbitals and, 42–43, 45, 46, 56, 59, 162n6; in oxidative stress, 165n8; in phosphate groups, 108–9, 164n1;

INDEX | 189

electrons (*continued*)
pi bonds, 46; polar covalent bonds, 48; properties of, 41–45, 56; protons and, 37, 39–45, 56, 110; removal of during cellular respiration, 103, 111, 113; resonance, 47, 108; role of in an atom, 39, 40, 41–45, 110; role of in chemical reactions, 40, 45; sharing of between atoms, 45–47; shells and, 42–47, 49–50, 59; sigma bonds, 46; significance of electron energy to living matter, 41, 49, 87–88; spinning of, 40, 42–43, 44, 45, 56, 59, 162n6; theft or stealing of between atoms, 44, 45, 47, 48, 49

elements, table of, 37

embodiment: acknowledgement of its by an organism, 3, 11, 25–26, 31, 105, 148, 149–50; as a construct to facilitate twentieth-century scientific research, 54–55. *See also* materiality

endocrine system: chemical disruption of the, 157; nerve connection of autonomic system to the, 74, 148, 163n9; organs and functions of the, 74

energy: atomic elements as conglomerations of, 32–33; as basis for response of an organism to its environment, 70, 75, 144, 148, 150; bodies as energy-in-transition, 28, 29, 105–6, 116, 148, 150; bonding in transformation of, 56–57; breathing, 103–4, 106–8, 113–16; carbohydrate molecules, 111–12; carbon-based, 49–50; cells and cell membrane porosity as product of self-constraint of, 51–52, 53, 56–57, 68, 70, 75, 101; cells, cell membrane porosity, and cellular activity as energy-in-transition, 106–10, 116–17, 119–20, 144; cellular respiration as energy-in-transition, 103, 105–6, 108, 110, 111, 115–18; characteristics of atoms as product of electron-proton interactions, 41, 110; characteristics of molecules as product of electron-proton interactions, 41, 45–46, 51–52, 55–57, 104; chemical and biochemical reactions as products of balances and imbalances of, 40–41, 45, 55–57, 102, 104; covalent and polar covalent bonds, 45–46, 47, 48, 50; diffusion as a transformation of, 55–59, 102, 103, 107; as dynamic play of in/determinacy, 34–35, 39–40; eating, 102–4, 107, 110, 111–12; electrons as non-sequential, shifting flows of negative, 39–42; energy-in-transition versus stuff-built-or-composed, 104–5, 110, 116; imbalance of between protons and electrons, 40–44; imbalances of, 38; in fatty acid molecules, 59, 111; in water molecules, 59–60; matter as energy: example of magnets, 34, 35, 36; metabolism as appropriation of, 110–16; mitochondria, 113–14; molecular influx and efflux through cells as a transition in energy, 101–2, 108–10, 119–20, 163n7, 164n2; movement of within an atom, 35–36, 40–42; neutrons as particle-like forms of, 36; nuclear reactions, 37–38; nucleus of an atom as a zone or field of, 36; organic matter as a product of self-constraint of, 33, 40–41, 49–52, 119, 148; oxygen and transitions-in-energy of organisms, 105–7, 111, 113–18; "pieces" of an atom as forms of, 35–36; precision of energy underlying formation of matter, 33, 51, 85, 111, 163n7; protons as forms of positive, 36; quarks as forms of, 36, 38; resonance, 47, 66, 108, 109, 164n1; self-constraint or self-relation of as basis for formation of matter, 25–27, 33–37, 45, 51–52, 53, 70, 119; stable form of matter as product of self-organizing capacity of, 31, 51–52; states or levels of, 57, 59, 60, 62, 107–9; tension and constrained vibration in transformation of, 56–57, 107; transitions in, 101–2, 104–7, 109–10, 116, 118, 119–20, 122; transitions-in-energy as link between an organism and its environment, 106–7,

190 | INDEX

116–18, 119–20, 144. *See also* matter; substance
environment: activities as bases for distinguishing a cell from its, 68, 75–76; activities as bases for distinguishing an organism from its, 75–76, 82; as a factor in evolutionary change, 135–36, 138, 140–44; Alaimo on transcorporeality of an organism and its, 67–68; climate change, 2, 12–13, 153; as condition of possibility for development of an organism, 4, 11–13, 17–18, 20, 68, 85, 118; dangers of engineering interactions between organisms and their, 155–58; developmental plasticity of organisms as responses to, 132–36, 138–44; environmental degradation, 2, 12–13; epigenetic responses of organisms to, 125, 126–27, 135–36, 139–44, 165–66n2; genetic influences on organic reactions to, 91, 123–34, 134–43, 165–66n2; genetic research on organisms and, 16–17, 54–55, 77–78, 125–26, 130–31, 135–38; hormones and steroids as influences on an organism's interaction with its, 74–75, 91, 139; interdependence of organisms and their, 26, 28–29, 39, 102, 128–29, 164n7; "itness" of an organism that separates it from an, 121–23, 134, 145, 148; materiality of as an influence on an organism, 20–21; molecular and cellular activities as responses to, 55, 69; nerve cells as means through which an organism interacts with its, 70; oxygen-based facilitation of interactions between organisms and their, 106, 117–18; porosity of an organism as its connection to an, 17–21, 52, 55, 102, 119, 124–25, 130; protein production as basis for organic reactions to, 91, 97, 125, 165–66n2; reproductive toxicology, 127–28; responsive basis for an organism's survival in an, 89, 90–91, 97, 119–24, 126–34, 134–45; sociocultural influences of on organisms, 17, 21, 76, 151–58; stress in an, 3, 133, 139, 155; temperature, 60, 64–65, 73, 74, 130; temporal layering of an organism's response to, 144–45; toxins, 12, 126, 127, 128, 129, 133, 157. *See also* habitat

enzymes, 79, 108, 112–13, 128–29, 165n6

epigenetics: debunking of myth of master molecule, 82–83; epigenetic biopolitics, 155–57; epigenetic factors, 82; epigenetic marker management of gene transcription, 96–97, 125, 140, 142, 165–66n2; epigenetic markers as an influence through generations of organisms, 29, 96, 135–36, 139, 141, 142–44; epigenetic molecules and biomolecules, 139–40, 144; methyl groups, 140–41, 164n6; response to temporal and environmental factors by organisms as a concern of, 82, 124–25, 135–36, 141, 153–54; transgenerational change, 123, 135–36

evolution: as an organism's response to material and social changes in its habitat, 142; concept of mutation-driven, 135; concept of sequestered, 5, 81, 98, 124, 126, 126–27; cross-generational transformation, 29, 134–35, 144; Darwinian theory, 54; deterministic and undeterministic theories of, 81; developmental plasticity a contributor to cross-generational of organisms, 134; genes as vessels containing recipes for an organism's response to its environment, 91; reproductive toxicology, 127–28; resistance of some proteins to, 143; scholarly debates concerning of organisms, 135–36, 166n3; substantive theory of, 135; theory of environment-driven, 135–36, 142–44. *See also* developmental plasticity

fatty acids: covalent bonds in, 59, 162n3; saturated, unsaturated, and polyunsaturated fat, 162n3. *See also* lipid molecules

Fausto-Sterling, Anne, 14–15, 17, 120, 157

fertilization: development of organisms from fertilized eggs, 125; eggs, 124, 127–28, 132; influence of environment on development of fertilized germ cells, 130, 139–40; influence of RNA molecules during, 131–32; meiosis, 130; mitosis, 130, 131; myth of impotency of abnormal sperm cells, 128; pluri-potential state of just-fertilized germ cells, 126; process of, 128; role of regulatory genes, 131–32; sperm cells, 127–28

fluid, 60, 61

food: as a means of supporting bodily dismantling processes, 102; breaking-down-of-food reactions, 111; cellular respiration of, 103, 104; common conceptions of bodily uses of, 78, 102; glycolysis, 111–13; influence of on regulatory genes, 133; ingestion of to support bodily energy transitions, 107, 110, 118, 122; link between eating and breathing, 103, 104; transitions in energy versus consumption of, 28, 102–3, 104–5; use of to regenerate ATP molecules, 110. *See also* eating

Franklin, Sarah, 81

Garrett, Reginald H., 78, 79, 88–89, 109–10

gas, 17, 41, 103, 105, 115

gates: in cell membranes, 54, 65–67, 69–70, 74–75, 78, 79, 90; cell transporter, 69, 74, 108; molecular in DNA, 93, 97–98; in nerve cells, 108–9, 165–66n2; protein, 75, 97, 108–9; sodium-potassium in nerve cell membranes, 71. *See also* channels; pockets; pores; transporters

generations: carriage of epigenetic markers on DNA through multiple of organisms, 29, 82, 91, 96, 136, 142; carriage of response-to-habitat through multiple of organisms, 29, 82, 122–27, 129, 134, 136, 139–40; cellular porosity as basis for development and transformation of multiple of organisms, 124, 129, 134, 145; constrained self-relation of energy as an influence on multiple of organisms, 35; cross-generational templates and transformation, 29, 134–35, 144–45; developmental plasticity as a factor in cross-generational change, 132–44; genes that escape epigenetic programming through multiple, 142–43; influence of germ cell methyl groups on multiple of organisms, 140–41; intergenerational openness to culturing of biocultural creatures, 156; intergenerational time: example of Russian doll, 134; itness of organisms as a product of intergenerational response to habitat, 134, 145, 151; manifestation of genetic mutation through multiple of organisms, 138; non-contemporaneity of an organism's habitat-responsiveness, 145, 151; population-sized differentiation over several, 141–42; production of epigenetic markers that can be carried through, 139–40; transgenerational change, 122–23, 135

genes: alleles, 137, 165–66n2; biocultural nature of, 77, 98–99, 101; concept of sequestered, 81, 98, 124, 126–27; constraints and conditions of possibilities underlying biochemical activities of, 83–84; developmental plasticity, 132–44; epigenetic factors and markers, 82, 94, 96–97, 125, 139, 140–44, 165–66n2; fiction of genes as being things, 82; function of as nodes of durable action, 82; function of as recipes, reference points, or codes for making proteins, 91, 94, 143, 163n11; gene expression, 143–44; genetic codes: an example using letters of alphabet, 95–96; genetic codes, 94–95, 137; genetic sequences, 1, 95–97, 137, 141, 143, 165–66n2; genetic transcription, 96–98, 125, 138–40, 165–66n2; habitat as an influence on selection and manifestation of in an organism, 17, 54–55, 82–85, 98, 101, 129–33, 139; heterochrony, 133; junk DNA codes, 94; morphotypes, 132–33, 145; mutation, 54, 126, 135–39, 142, 143, 144, 166n3; number of in human genome, 94; presence of shared among different species, 8, 142–43; random mutation, 135, 137, 138, 143; regulatory, 96, 131–33, 137, 138; role of in making proteins, 80, 83–84, 91, 94, 96, 125, 165–66n2; role of in regulating and facilitating cellular traffic, 27; somatic mosaicism, 97

genetics: breeding and interbreeding, 8, 54, 135; Darwinian theory, 54; early twentieth-century ideas regarding genes, 81–82; example of stickleback fish, 133; fiction of genes as being things, 82; fiction of genes as master molecules, 82–83, 105; fiction of sequestered genes, 81, 98, 124–27; fiction of directive or causative effects of genes, 83, 91; fiction of überbiological or übernatural genes, 80–85, 124, 126; habitat as a determinant of gene selection and manifestation, 17, 54–55, 132–33; Keller on early twentieth-century, 54–55, 77–78, 81–83, 94; presence of shared genes among different species, 8; research in, 16–17, 54–55, 81–83, 124–25, 139, 143

genomes, 8, 94, 98, 144, 163n11, 166n3

germ cells, 29, 125–33, 134–37, 138–42, 144

gestation, 97, 127, 132, 134, 139

glucose, 69–70, 112–13, 115, 117–18, 164n4. *See also* carbohydrates

glycolysis, 111, 112–13. *See also* citric acid cycle; Krebs cycle

gradients: cellular increase or enhancement of concentration, 107–9; chemical, 71, 163n8; co-transport as a means of compelling molecules against concentration, 67; diffusion of molecules down concentration, 57, 72; electrical, 71; electrochemical of mitochondrial membranes, 114–15; formation of electrochemical to accommodate ATP, 111, 115; glucose molecule travel down concentration, 112; hydrogen ion travel up concentration, 114–15; movement of oxygen in concentration gradients during breathing, 115–16; reduction of concentration, 162n2; role of transporters in creating and maintaining electrochemical, 163n11. *See also* concentration

Grisham, Charles M., 78, 79, 88–89, 109–10

Grosz, Elizabeth, 7–8, 25, 26

guanine nucleotides, 93, 141

Guerrero-Bosagna, Carlos, 139–41, 144, 156, 166n3

Gunnarsson, Lena, 51

Guthman, Julie, 17–18, 20–21, 157

Habermas, Jürgen, 5–6, 10

habitat: biochemical connections between an organism and its, 28–29, 117–18, 129–30, 151–52; biocultural habitats, 152–59; dangers of ignoring sociocultural impact of, 153–55, 157–59; dangers of manipulating sociocultural habitats, 155–57; developmental plasticity as a carryover through generations of organisms, 132–33, 134–40, 140–45, 148–52; differentiation and divergence in populations, 140–42; differentiation versus reproductive isolation, 140–42; embeddedness of an organism in its, 27, 54, 75–76, 119, 122, 151–52; function of as condition of restraint or possibility for an organism, 118, 124, 129, 150–51; habitat-induced heterochrony, 133; influence of on changes in DNA of an organism, 140–42; influence of on genetic responses of an organism, 84, 98–99, 101, 122–24, 129–33, 134–40, 144–45; influence of on transitions of energy of an organism, 106, 117–18, 119, 144–45; interdependence of organisms and their, 4–5, 17, 25, 28–29, 151–52; myth of genetic sequestering of organisms from their, 124, 126–27, 142, 148–49;

habitat (*continued*)
noncontemporaneity of an organism with its, 29, 121–24, 133–34, 144–45, 149–52; organism-that-responds versus organism-as-response, 134–40, 148–52, 165–66n2; perception of and response to by an organism, 75–76, 91, 98–99, 101, 122–24, 144–45, 148–52; sociocultural influences of on a biocultural creature, 17, 20, 133, 152–59, 165–66n2; speciation as an organic response to, 140–42. *See also* environment

Hall, Judith G., 97
Hancock, Ange-Marie, 154
Haraway, Donna, 81, 82–83
helium, 37
helix, 87–88, 93, 141
Hemming, Clare, 21
heterochrony, 133
Hibbing, John, 153–54
Hill, Christopher, 42, 161n1, 162n6
Hird, Myra, 25, 102, 163n6, 165n1
Hobbes, Thomas, 21–22
Hood, Ernie, 157
hormones: hormonal cascades, 91, 128–29; influence of on germ cell porosity, 128–29, 131, 132; movement of through cell membranes, 65, 74; nonpolarity of, 65; production of by endocrine organs, 74, 157; regulatory functions of, 74–75, 163n10; research on by Celia Roberts, 13–14; role of cellular in interaction of an organism with its environment, 91, 125, 131, 132, 139, 148
human exceptionalism, 1, 10, 13, 147, 158
humans. *See* animality of human beings; biocultural creatures; organisms
hydrogen: absence of neutrons in, 38; as a building block of life, 49; atoms in a carbohydrate, 111, 112; atoms in a fatty acid, 59, 110, 111, 162n3; atoms in an amino acid, 86; atoms in a water molecule, 59–61; atoms in methyl groups, 140; in breaking down of food, 102, 110, 111; hydrogen bonds, 45, 48–49, 60–61, 86; hydrogen ions, 114–15; in Krebs cycle, 113–15; in metabolization, 102, 110, 111; oxygen-hydrogen appendages in RNA, 92; protons in, 37; in pyruvate molecules, 112–13, 115; removal of hydrogen ions through breathing, 103; table of elements, 37

hydrophobic and hydrophilic interactions: analogy of oil and vinegar interaction, 62; clustering phenomenon of, 62; formation of membranes through, 62, 64; hydrophilic molecular heads, 62–63; hydrophobic molecular tails, 63; micelles, 63; polarity and nonpolarity in, 62–63, 65, 162n4

ice, 60
identity, 20, 26–27, 52, 148, 151–53, 157
imbalance: energetic imbalances on either side of a cell membrane, 55, 59, 108; influence of an internal imbalance on interaction of atoms, 43–44; an between protons and electrons, 43, 44; an between protons and neutrons, 38
Inda, Jonathan Xavier, 14, 155
influx and efflux: cell membrane facilitation of chemical, 27, 65, 68–69, 76, 101; as changes in energy rather than of substances, 101; and changes in gradients, 71; concepts of inside and outside, 54, 67, 101–2; discretionary nature of cellular, 52, 68–69, 70–71, 75–76, 98, 135, 150; importance of to continued biochemical activities, 67, 75–76, 90, 107, 131; as part of an organism's activity-of-response to its environment, 122, 131–32, 135, 150; three-dimensional quality of, 68. *See also* cell membranes; permeability; porosity

ingestion, 28, 104, 107, 122, 145, 148
inside and outside: activity as basis for demarcating of a cell, 27, 53–55, 68, 71, 75–76, 77; biochemical activity as basis for defining, 77, 104, 124; influx and efflux of molecules through cell mem-

branes, 54, 58, 68–69, 83; molecular diffusion reactions, 64, 71–72; substantive conceptualizations of, 27, 54, 67, 77, 104, 121, 124, 135. *See also* boundaries
ions: anion, 47; atomic formation of, 47; cation, 47; chloride, 60–61, 163n8; difficulty in constructing from carbon atoms, 50; hydrogen, 103, 111, 113–15, 165n8; ionic bonds, 47–48; iron, 115, 165n7; negatively charged, 47; positively charged, 47, 71; potassium, 70–72, 108, 109; sodium, 66, 71–73, 143; transport of by cell membranes, 71–74, 108

Keller, Evelyn Fox, 18, 54–55, 77–78, 81–82, 94, 98, 124–25
Klug, William S., 131, 132, 133, 143, 164n4
knowledge. *See* politics
Koeppen, Bruce M., 74, 164n3
Kohn, Eduardo, 152
Krause, Sharon, 11
Krebs cycle: process of citric acid or, 113–16; pyruvate molecules in the, 112–13, 115. *See also* glycolysis
Kristensen, Anders R., 90
Kültz, Dietmar, 102
Kuzawa, Christopher W., 155

Landecker, Hannah, 54, 134
Lane, Nick, 91–92, 165n5
Latour, Bruno, 153
Lederman, Leon M., 42, 161n1, 162n6
Lewontin, Richard, 82, 88
lipid molecules: in cell membranes, 59, 63–66, 110–11; lipid bilayer membranes, 63; movement of, 64–65. *See also* fatty acids
liquid: cellular transport, 79; diffusion in, 162n2; formation of by atoms, 41; liquid water, 60; molecular, 58–59
Lock, Margaret, 148–49
Lodish, Harvey, 88, 112–13, 115, 162n2, 164–65n4, 165nn7–8
Lordkipanidze, David, 144
Lupski, James R., 97

McEwen, Bruce, 75–76
Mansfield, Becky, 17–18, 20–21, 157
Marder, Michael, 25
Martin, Emily, 127
Mashoodh, Rahia, 156
master molecule, 82–83, 98–99, 105. *See also* überbiological matter
materiality: as a theoretical basis for opponents of human exceptionalism, 20, 32–33, 82, 148–49; as a theoretical basis for proponents of human exceptionalism, 13, 19, 54–55, 84–85, 147; criticisms of assumptions underlying human, 11, 13, 18–19; embedding of environment into a body, 20; materialist outlook of Hobbes, 21–22; materialist understanding of self, 22, 25–26; of language, 21–22; shared of all things, 32–33, 82, 148–49. *See also* embodiment
matter: atomic elements as conglomerations of energy rather than as "stuff," 32–33; bonding of atoms to form, 45, 56; carbon as basis for organic, 33, 49–52; chemical reactions between bonding atoms as a means of transforming energy into, 56–57, 58, 64; diffusion as a means of transforming energy into, 58; formation of through constrained self-relation of energy, 25–27, 33, 35–37, 45, 49–52, 70, 119; influence of electron-proton reactions on creation of, 40–41, 55–56; matter as energy: example of magnets, 34, 35, 36; myth of genes as überbiological or übernatural matter of life, 81–83, 98–99; precision underlying formation of, 33, 51–52, 85; as product of conditions of possibility, 32, 53, 148, 151; proton count as an influence on identity of, 37; self-organizing nature of, 31–32, 51, 55–56, 107; as substance, 33; substantialist thinking, 33–34. *See also* energy; substance
Mayr, Ernst, 126
Mbembe, Achille, 155–56
meiosis, 130, 132

Meloni, Maurizio, 16, 149, 155
membranes. *See* cell membranes
Mendenhall, Ruby, 155
metabolism, 103, 110–16, 118. *See also* cellular respiration
methyl groups, 140–41, 164n6
micelles, 63
mitochondria, 113–14, 165n5
mitosis, 130–31
molecules: adenosine, 164n2; amino acids, peptides, and polypeptides, 85–88, 94–95, 98; amphipathic, 62–63, 64; atomic elements basic to formation of organic, 49, 51; ATP, 109–13, 115, 164n2; biochemical, 85–91, 96, 104, 134–35, 142–43; bonding effects, 59–61, 107, 110; carbohydrate, 65, 110, 111; carbon, 49, 50–51, 59, 112; carbon dioxide, 65, 102, 115–16; cell membrane: analogy of ping pong balls in bathwater, 64; cell membrane, 53–55, 58–69, 75–76, 78, 90–91, 98, 101; chemical reactions among, 58, 64–65, 68–69, 85–91, 95–98, 104, 107; cholesterol, 64; concentrations and concentration gradients, 57, 64, 65, 67, 70, 90, 107–10; cotransport of, 66–67; diffusion, 57–58, 59, 63, 64–67, 69, 112, 150; distribution of disordered, 57–58; DNA, 92–98, 131–32, 140; electron-proton tensions of atoms underlying molecular porosity, 41, 51–52, 56, 110; endocrine, 73; energetic, 49; energy and high-energy, 103, 104, 108; epigenetic factors and markers, 82, 96, 140; fatty acid (lipid), 59, 64, 65, 110, 118; genes, 94–98, 131–32, 140; glucose, 69–70, 110, 112, 115; hormone and steroid, 74; huddling and jostling action of membrane, 59; hydrophilic interactions, 62; hydrophobic interactions: analogy of oil and vinegar, 62; hydrophobic interactions, 61–63, 64, 162n4; influence of atom bonding on shapes of, 46, 47–49, 51, 56–57, 59; influence of atomic patterns of resonance on, 47; influx and efflux of through cell membranes, 52, 54, 65, 67, 75, 90, 98, 101–2; interaction of energetic with inactive molecules: analogy of individuals holding hands moving through a crowd, 61–62; interaction of energetic with inactive molecules, 59, 61; interaction of vibrating, 56–57; introduction of into an organism from environment, 20–21; large, 65; linear chain formations by, 59; lipid bilayer membranes, 63; micelles, 62–63; mixture of amphipathic molecules with water: example of detergent, 62; mixture of amphipathic molecules with water, 62–63; mixture of oil molecules (fatty acid) with water, 61–62; mixture of salt molecules with water, 60–61; molecular activity as energy-in-transition, 104, 106; molecular memory, 92–93; molecular traffic, 54, 65–74, 78–79, 90, 101, 106–7, 165–66n2; molecular traffic speeds, 67; movement of, 56–58; movement of targeted across cell membranes, 65–66, 112–15, 163n7; myth of master or überbiological, 82–83, 98–99, 105; nerve cell, 70; neurotransmitters, 73; nonpolar, 61–62, 65; oxygen, 165nn8–9; phosphate group, 108–11, 115; phosphorylation, 108–10; polarity in interaction of, 61–63, 65–66; precision of molecular processes, 27–28, 51, 79–80, 85–90, 93–95, 150; protein, 64, 65, 78, 85–97, 108, 113, 132; pyruvate, 112–13, 115; reactive interactions among, 85–91, 94, 96, 107–8, 111, 117, 165n8; reactive interactions with an organism's environment, 117, 120–22, 131, 134–35, 140–43, 150, 161n1; repulsion between, 57; resonance, 47, 66, 108–9, 164n1, 165n7; reverse transcription, 96–97; RNA (ribonucleic acid), 91–92, 98, 131–32; self-constraint of energy as an influence on conditions of possibility of, 51–52, 86–90, 104, 110, 150; self-sorting of, 58; shapes of

as determinants of their interactive capacities, 46; shifting of energy states by, 56–57; signaling, 98, 164n6; small, 65; stable, quiescent, chemically inactive, 59; tagging of, 69–70; transformation of into matter, 58; vibration of, 56–61, 73, 161n1; water, 59–61, 66, 78, 162n2
morphology, 132, 133, 141, 157
Morris, David B., 152, 153
Morton, Timothy, 12, 32, 52
mutation, 54, 126, 135–38, 142, 143, 144, 166n3
Mychasiuk, R., 133

negative charge: as a key factor in chemical reactions, 40, 165n8; as a shifting flow of negative energy, 39, 41, 42, 56; cations and anions, 47; imbalance of electrons with positively charged protons in an atom, 41–43, 56, 110; in a water molecule, 60; influence of on gradient of cell membranes, 71–72, 163n8; influence of resonance, 47, 164n1, 165n7; in polar covalent bonds, 48; of electrons, 39, 40, 41; of some atoms, 44; oxidative stress in molecules, 165n8; self-interaction of negative energy in electrons, 39–40. *See also* charge; positive charge
Nelson, David L., 65, 67, 68, 78, 90, 109, 162n4
nerve cells: autonomic, 74; axons, 70, 73; dendrites, 70, 73; depolarization and repolarization in, 72–73; as mechanisms that enable an organism to absorb and respond to its environment, 70–71, 73–75; movement of ions underlying change in charges of, 70–73; neurotransmitters, 73, 165–66n2; of multicellular organisms, 73; permeability of as basis for their perception and response to environs, 71, 109–10; physical configurations of, 70; serotonin, 165–66n2; sodium-potassium gates in membranes of, 70–72, 108, 109; synapses, 73
neutrons: composition of, 36; nuclear reaction, 161–62n5; nuclear transmutation, 38; properties of, 36, 38; protons and, 36, 38, 161–62n5; quarks in, 36, 38; rogue, 38; "strong force" of, 36; transformation of into protons, 38
Nichols, John G., 72–73, 163n8
Niewöhner, Jörg, 148–49
nitrogen, 38, 49, 65, 80, 110–11
norms, assumptions, and expectations: biological embedding of social perceptions, 149; dominance of human exceptionalism ideas over, 10, 11–12; as factors that shape biocultural creatures, 20, 21, 23, 25–26, 120–21, 148–55; influence of hegemonic political formations over, 157; influence of over scientific research, 14–15, 124; persistence of outdated among scientists, 18–19, 33, 124–25, 128; persistence of outdated among social scientists, 82–83; redefining of human as a means of consciously addressing and questioning, 10, 19–21, 31, 77, 120–21, 148–49, 156–57; traditions of exclusion and inclusion of specific social groups, 9, 127, 157–58. *See also* politics
Nosil, Patrik, 138, 140
N-terminus, 86–87. *See also* C-terminus
nuclear reactions, 38, 161n3, 161–62n5
nucleotides: adenine dinucleotides, 165n6; changes to through genetic copying errors, 136; four basic in DNA, 93, 94–95; pairing preferences of four basic, 141; replicator assemblages, 94; ribonucleotides, 92, 141
nucleus: activity around a, 39, 41; activity in a, 36–37, 41; changes in composition of an atomic, 37–38, 161n5; electrons around a, 39, 41–45, 56; in covalent bonding, 48; in ionic bonding, 47; of a carbon atom, 50; of a cell, 69, 79, 93–94, 97, 139, 164n7, 165n1; of an atom, 36; of an oxygen atom, 60; protons and neutrons in a, 36–37, 41, 44; "strong force" in a, 36

INDEX | 197

ontology: biocultural approach as a reorientation of humanity to world, 19; embodied subjectivity, 20, 23, 25–26, 31, 121, 153; human as a posthumanist category, 24; inherently political nature of biocultural humans, 159; itness of an organism, 121, 134, 145; ontologically indeterminant nature of matter, 51; posthumanist, 158–59; Renaissance valuation of beings based on their political dignity and authority, 158–59; substantialist, 33–34

orbitals: in atoms, 42, 45; in chemical bonding, 45–47; electron in carbon, 49–50; electron in oxygen, 59–60; electrons and, 42–43; molecular, 45–46, 56, 110–11; properties of, 43; shells, 42–43

organisms: activities as basis for distinguishing "inside" from "outside" in cells and in, 75–76, 150; as a gross form of self-relation of energy, 41; animality of human beings, 3, 4, 13; biochemical processes that enable multicellular entities to function as, 70, 75–76, 89–91, 94, 109, 117–18, 129–30; biocultural nature of living, 27–29, 76, 133–34, 147–48, 150; carbon balance between and their habitats, 38–39, 150; cell membrane permeability as means of enabling life processes in, 55, 70, 75–76, 83–85, 89–91, 121–22, 124–35; chemical bonding as basis for life processes in, 45; deaths of, 39; developmental plasticity in, 132–33, 134–36, 138–44; DNA coding and response-to-habitat of, 82, 94–98, 125–26, 130, 136–37, 140–42; early twentieth-century genetic theory regarding coming-about of, 81–82; epigenetic responses of, 82, 125, 136, 139–45; evolutionary change, 140–44; genetic activity as response of to their habitats, 82–85, 90–91, 94–99, 122–24, 129–40, 144–45, 163n11; germ and somatic cells of, 125–26, 127–28, 129–32, 134, 138–42; habitat as condition of possibility for, 119, 122–24, 133–34, 136, 138–39, 150; as histories-of-response, 123–24, 130, 136, 150–51; interaction of with environment as basis for their development, 82, 84, 89–90, 94, 101–2, 134–35; loss of equilibrium by with their habitats, 39; mutations in, 126, 135–37, 137–38, 142–43; myth of überbiological matter or molecules, 80–85, 98–99, 120–21, 124, 126; nerve cells function as an interface between organisms and their habitats, 70, 73–74, 75; organism-that-responds versus organism-as-response, 134–40, 148–52, 165–66n2; oxygen as an interface between organisms and their habitats, 102, 103, 105–7, 117–18; perception of environment by, 75, 91, 99, 109, 122, 129, 145; precision of biochemical processes of, 27–29, 84–87, 89–99, 120, 122, 164n4; reaction and diffusion processes of, 58, 70; rebuilding function of multicellular in response to environmental stimuli, 75–76, 120, 134, 144–45; responsiveness of to their habitats, 89–91, 94, 98–99, 119–24, 129–36, 138–45, 148–55; scientific rethinking of interaction of with environment, 16–18, 23–26, 80–81, 124, 129–30, 142–43; self-constraints of energy that give stability to, 51, 70; stability of some proteins across different species of, 143; stable self-constraints of energy that determine conditions of possibility for life for, 51; temporal nature of relationships of with their habitats, 29, 99, 121–24, 134, 144–45; tensions of energy contained in, 53; transitions in energy between organisms and their habitats, 104–7, 110, 118, 119–22, 150; zoosemiosis, 163n3

osmosis, 162n2

outside. *See* inside and outside

oxygen: as an indicator of an organism's engagement with its habitat, 28, 106–7, 118; aspiration of air, 28, 104; atoms in carbon dioxide, 115–16; carbon monoxide

displacement of in blood, 165n9; carbon-oxygen atoms in polypeptides, 86; cyanide interruption of electron transport chain, 115; function of in an electron transport chain, 113–14, 115; function of in breaking down food to produce energy, 103, 107, 108–11, 113–14, 116–17; function of in cellular respiration, 103, 105–6, 110, 111, 117; function of in Krebs cycle, 113; function of in waste disposal processes of organisms, 103, 106; importance of to functioning of organisms, 28; in a phosphate group, 164n1; in blood, 103, 115–16; in carbohydrate molecules, 111; in fatty acid molecules, 110–11; in water molecules, 59–60; metabolic function of in an organism, 103, 105–6, 110, 117; myths and misunderstandings regarding function of, 103–4, 105; oxidative stress, 165n8; oxygen atoms, 46, 49, 59–60, 115; oxygen-hydrogen appendages in RNA, 92; oxygen molecules, 65; phosphorylation, 108–9, 110; transience of molecules in an organism, 103–4, 105, 106–7, 116–18, 131
Oyama, Susan, 18, 23–24, 26, 84

Painter, R. C., 133, 139
Panofsky, Aaron, 134
Papoulias, Constantina, 120
peptides and polypeptides: chaperones, 164n4; peptide bonds, 86, 87; peptide folding, 88; polypeptide formation, 86–89, 137
permeability: and concepts of inside and outside, 53–55, 67–68; distinction of activities through cellular, 67–68; facilitation of cellular activity through, 27; linkage of an organism to its environment through, 17, 18–19, 55, 121–22, 124; as means through which entire organisms interact with their habitats, 70–71, 76, 85, 91, 101–2, 119–20; of atoms and of organic flesh, 41, 52; of germ cells as a basis for an organism's response to its habitat, 29, 126–27, 130, 135, 152; of nerve cells, 70–71; of human body to social and to environmental influences, 17, 83, 121–22; role of in transition of energy into materiality, 52, 55; selective nature of cell membranes, 65, 68–69, 85, 90, 101, 107, 119, 150. *See also* cell membranes; influx and efflux; porosity
phosphate: (GDP) guanosine carriers, 164n2; ATP, ADP, and AMP carriers, 164n2
phosphate group, 108–10, 111, 112, 114–15, 164nn1–2
phosphorus, 49, 110–11, 164n1
phosphorylation, 108–9
physics, 25, 26, 33, 161n1
Pinto-Correia, Clara, 82
plant life, 25, 38–39, 41, 78, 105
pockets: function of in a cell membrane, 66; function of in ATP creation, 114–15; function of in DNA copying, 93, 97–98; function of in polypeptide folding, 88; hemoglobin, 116; molecular complementarity mechanisms, 93. *See also* channels; gates; pores
polarity: amphipathic molecules, 62–63; depolarization, 72–73; molecules that are neither wholly polar nor nonpolar, 62–63; nonpolarity of fatty acid molecules, 61–62, 65; nonpolar molecules, 61; of carbon, 50; of magnetic energy, 34; of oxygen, 111, 114; of protons and electrons, 39; of water molecules, 60, 61, 65; repolarization, 73. *See also* charge
politics: biocultural concept of mankind as a mechanism through which to address world crises, 3, 20–21; challenge to change that is presented by dominant political norms, 14–15; collective and democratic approaches to world's problems, 14, 20–21, 157–59; concept and category of human, 1–12, 16–19, 24–25; culture and, 4; dangers of reductionist thinking, 153–54; environment and climate as actors in global, 13, 31; fiction of human exceptionalism, 1–2, 5, 8, 11;

politics (*continued*)
Hobbesian theory, 21–22; human responsibility for current world crises, 1–2, 12–13; influence of hegemonic political formations, 157; knowledge production and authority, 14–17, 19, 23–24; management and manipulation of populations and of specific social groups, 14, 91, 153–58; materialist approach to, 21–22, 32; necropolitics, 155–56; neoliberalism, 14, 21, 153–54; political action, 14; political nature of sciences, 14–16, 153–54; political nature of social sciences, 14, 15–16, 83, 153–54; subjectivity, 4, 20, 23, 25–26, 31, 121, 153–54. *See also* biopolitics; norms, assumptions, and expectations

pores: aquaporin, 66; in cell membranes, 54, 66–67, 69, 71, 74, 78, 90. *See also* channels; gates; pockets

porosity: cell membrane, 27–28, 68, 77, 89, 125, 150; and concepts of inside and outside, 53–55, 75, 77; contribution of to cellular responsiveness, 55, 68, 75–76; diffusion and, 55, 58, 64; facilitation of molecular traffic through, 55, 64, 65, 68; germ cell, 128–29, 134–35, 138, 145; linkage of an organism to its environment through, 106–7, 121–25, 128–29, 134–35, 138, 145, 151; of human body, 17–18, 29, 106–7, 123, 125; produced by atoms and molecules, 51–52; self-relation of energy as matter, 26, 75; viscous, 18. *See also* cell membranes; influx and efflux; permeability

positive charge: as basis for movement of sodium and potassium ions in nerve cells, 71–72; cations and anions, 47; imbalance of protons with negatively charged electrons in an, 41–43, 110; in a water molecule, 60; influence of resonance, 47; in polar covalent bonds, 48; of hydrogen ions in a respiratory (electron transport) chain, 113–14; of protons, 36–37, 41; of some atoms, 44; proton-based identity of elements, 37. *See also* charge; negative charge

posthumanism, 24, 25, 158–59

potassium: influence of endocrine biochemicals on cellular, 74; potassium ion channels in nerve cells, 70–71, 71–72; sodium-potassium ATP-ase, 163n7; sodium-potassium gates in nerve cell membranes, 71, 108, 109

precision: enabling, delimiting and directional quality of biochemical, 27, 28, 64, 79, 84–85, 89, 120; function of at molecular level, 83–94, 84–90; genetic, 91, 93–94, 141; of cell membranes as determinants of cellular responsiveness, 70; of energy underlying composition of a protein, 85, 91; polypeptide formation by amino acids as an example of molecular, 87. *See also* constraints

pregnancy, 97, 127, 132, 134, 139

promoters, 98, 165–66n2

proteins: amino acids as building blocks of, 85–88, 92–93, 95, 97–98, 164n7; as a percentage of cellular makeup, 78; biochemical conditions bathing and diffusing cellular that influence their activities, 85, 88–89, 91, 94, 97, 122–23; cellular construction and dissolution of, 75, 90–91, 98, 164n5; cellular structure: analogy of Los Angeles, 78–79; chaperones, 164n4; commonality in protein usage across species, 8, 143; composition as enabling and delimiting factor underlying activities of, 85, 125, 142–43; configuration of: analogy of bunched string, 87; DNA as memory or recipe for construction of, 92–94, 95–98, 125, 137, 139; enzymes, 79, 108; epigenetic markers, 96–97, 125, 139–40, 142–43; gene activity in constructing from molecules, 80, 83–85, 93–97, 140; gene activity in maintaining life of an organism, 91, 98, 123, 133; in gene mutation, 137–38, 139, 164n4; germ cell, 123, 126, 128–29, 131–32, 144–45; habitat as a key influence on an organism's use of bodily, 84–85, 89–91, 94, 97, 98–99, 122–23, 134–35; high-energy molecules and, 108–9, 111, 113–15; in cell membranes, 27, 65–66, 75, 78–79, 90, 108, 163n11; in electron transport chains, 113–14; influence of functional amino acid groups on functions of individual, 88; in Krebs cycle, 113–15; methyl group influences on, 140–41; mitochondrial, 113–14; multiple roles played by protein molecules in cellular function, 78–79; phosphorylation, 108–9; polypeptide folding to produce, 87–88, 98, 164n4; precision in functioning of, 79–80, 83–90, 94; protein structure and topography, 88–89; reactive interactions of with other molecules, 88–91, 129, 134; regulatory gene, 131, 132, 133; replication and replicators, 93–94; resonance patterns, 108–9; ribosomes, 97–98, 141; RNA activity to construct, 91–92, 93, 94, 97, 98; role of hydrogen bonds in folding molecules into, 49; shapes of membrane to accommodate molecular traffic, 65, 78, 88–89; somatic cell, 125, 129; transporters, 69–70, 108, 163n11, 164–65n4, 165–66n2

protons: as basis for distinguishing elements and substances, 36–37; electrons and, 39–45, 49, 56; in a phosphorylation group, 108, 164n1; in carbon, 49–50; influence of on bonding, 45–48; influence of on shell formation in atoms, 42–44, 49; influence of on charge of an atom, 44; influence of on interactions between atoms, 44–45, 46, 49; in oxygen molecules of water, 60; neutrons and, 36, 38, 161–62n5; nuclear transmutation, 38; number of in an atom, 36–37; as part of binding force underlying atoms in molecules, 110; positive charge of, 36; properties of, 36–37; quarks in, 36, 38; role of in an atom, 40; significance of proton energy to living matter, 41, 49–50; transformation of into neutrons, 38

pyruvate molecules, 112–13, 115

quantum physics, 25, 26, 34–35, 39–40, 51, 161n1

quarks, 36, 38

Rabinow, Paul, 16–17
race. *See* biopolitics
reductionism, 27, 84–85, 133–34, 149, 151, 153–54
regulatory genes, 96, 131–33, 137, 138
reproductive toxicology, 127–28
resonance, 47, 66, 108–9, 164n1, 165n7
ribosomes, 97–98, 141
Ribot, Jesse, 12–13
Richardson, Sarah, 156
RNA (ribonucleic acid), 92–93, 95–96, 97–98, 131–32, 137–38
Roberts, Celia, 13–14, 157
Roberts, Dorothy, 14–15, 155
Rose, Nikolas, 14–15, 155–56, 159
Rose, Stephen, 70
Rutter, Michael, 155

Saldhana, Arun, 155
salt (sodium chloride), 44, 47–48, 60–61
self-relation, 33, 35, 45, 49, 51, 70, 119
self-understanding, 7, 12, 16, 31, 148, 158
serotonin, 165–66n2
shells: anions and cations, 47; atoms containing complete, 43; atoms containing incomplete, 43–44; bonding, 45–46; characteristic of molecules in living matter, 49; chemical reactions, 45; electrons and, 42–45; formation of in atoms, 42; in carbon, 49–50; of atoms, 42–43; of oxygen atoms, 46, 59–60; of phosphorus atoms, 164n1; orbitals in, 42–43, 46; shapes of in atoms, 43
Skinner, Michael, 139, 156

INDEX | 201

Slavich, George, 17, 154
sodium: influence of endocrine biochemicals on cellular, 74; sodium chloride (salt), 44, 47–48, 60–61; sodium ion channels in nerve cells, 66, 70–73, 143; sodium-potassium ATP-ase, 163n7; sodium-potassium gates in nerve cell membranes, 71, 108, 109
somatic cells, 97, 125–29, 131, 135–36, 139–40, 144
species: commonality of genes and patterns and protein usage across, 8, 142; differentiation as a key factor in creating new, 135–36; divergence and differentiation, 142, 143–44; early theories of evolution, 37, 54, 81; early theories of sequestered development of, 135–36; epigenetic mechanisms as influences on changes, 139, 142, 143; example of stickleback fish, 132–33; genetic mutation within populations, 136–38; humans as a variegated hybrid of homo, 8; morphotypes, 132–33, 144; power of environment to induce transgenerational changes, 135, 139, 142; random genetic mutation, 135; speciation, 140, 142
Stanton, Bruce A., 74, 164n3
Steffen, Will, 2
steroids, 65, 74–75, 132, 163n10
stress: bodily response mechanisms to, 75, 126, 140; case of Afro-American women in urban Chicago, 155; case of white and upper-class women, 127; developmental and embryonic, 128; differential responses to among same species, 155, 165–66n2; environment-induced genetic heterochrony in response to, 133; oxidative, 165n8; serotonin, 165–66n2
strong force, 36
substance: absorption by organisms of substances from their habitats, 17, 20–21, 65, 67–68, 75, 83, 119–20; biochemical substances, 85, 89–90, 107, 129; "feel" of substance or matter as a product of a constrained flow of energy, 34–35; matter as, 33; organic reactions to substances in a habitat, 119–20, 129, 142; osmosis, 162n2; as product of self-constrained energy, 25, 33; protons as determinants of an atom, 36–37, 40–41, 44–45; substance as energy: example of magnets, 34, 35, 36; substance-oriented, inside-outside theories of evolution, 120, 121, 135, 145, 148; substantialist or substance-oriented thinking, 27, 32–34, 68, 75, 77–78, 101–5, 116. *See also* energy; matter
sulfur, 49
Sweet, Elizabeth, 155

table of elements, 37
Tanriverdi, F., 74–75
Tazuke, Sallie I., 128–29
temperature, 60, 64–65, 73, 74, 130
Testa, Giuseppe, 149, 155
thymine nucleotides, 93, 141
toxins, 12, 126, 127, 128, 129, 133, 157
transcorporeality, 18, 67–68
transcription: gene by epigenetic markers, 96; reverse-transcription, 96–97
transporters: allele production of, 165–66n2; glucose, 112, 164n4; influence of endocrine chemicals on, 74; number of in human genome, 163n11; role of in production and maintenance of electrochemical gradients, 163n11; sodium-potassium ATP-ase, 163n7; tagging of glucose molecules by, 69; work of in phosphorylation, 108; work of with DNA in response to chemical signals, 98. *See also* gates
Tuana, Nancy, 118, 153

überbiological matter: fiction of, 5, 98–99, 120–21, 145; fiction of genes as, 80–85, 124, 126. *See also* master molecule

Via, Sara, 127
Viau, V., 74–75
vibration: in bonding, 56; constrained,

57; formation of ice, 60; influence of on nerve cell response, 73, 161n1; intensity of, 56–57, 60, 61; molecular, 56–58, 60–61, 162n1; in polypeptide folding, 88

water: analogy of paired hand-holding in a crowd, 61–62; in an electron transport (respiratory) chain, 113–14, 115; in an endocrine system, 74; as an environmental concern, 10, 21; in a respiratory (electron transport) chain, 113–14, 115; behavior of molecules, 60; in carbohydrate molecules, 110–11; in cellular composition, 78; chemical notation for, 59; components of a molecule, 59–60; diffusion processes in, 162n2; diffusion process of, 162n2; in fatty acid molecules, 59; formation of ice, 60; formation of steam or vapor, 60; in genetic development, 131; hydrophobic interactions, 62–63, 162n4; interaction of amphipathic molecules with, 62–63; liquid water, 60; mixing of with oil, 61; mixing of with salt, 48, 60–61; properties of molecules, 60, 63, 65

Webster, Jeannette I., 75

West-Eberhard, Mary Jane, 130–31, 132–33, 138–39, 140, 141, 142, 143–44

White, Stephen, 159

White, Yvonne A. R., 128

Wilson, Elizabeth, 153, 165–66n2

Wittig, Monique, 21–22

Wolfe, Cary, 9, 11–12

Wolinsky, Howard, 138